Practical Radiography

Indian Society for Non-Destructive Testing – National Certification Board Series

Practical Radiography

Baldev Raj
B Venkataraman

Published for

Indian Society for
ISNT
Non-Destructive Testing

Alpha Science International Ltd.
Pangbourne, U.K.

Baldev Raj
B Venkataraman
Materials, Chemicals and Reprocessing Groups
Indira Gandhi Centre for Atomic Research
Kalpakkam, India

Alpha Science International Ltd
P.O. Box 4067, Pangbourne RG8 8UT, U.K.

ISBN 1-84265-188-9

Printed in India

Foreword

I am immensely happy and pleased to learn that the book titled "Practical Radiography" authored by Dr. Baldev Raj and Shri B. Venkataraman is now ready.

For effective execution of its various activities and for meeting some of its objectives such as Training and Certification of NDT/NDE personnel, the National Certification Board (NCB) felt it necessary to bring out appropriate books and reference material on various methods/ techniques of NDT. While some books on the subject dealing with the basics of NDT methods and techniques are available in the market, NCB however felt that practical guide books on various NDT techniques had to be published which not only provide information relevant to academic, research and industry but also comply with the requirements of National and International standards.

I am fully aware and conscious of the stupendous efforts that are needed in bringing out such a broad range of data/information with authority, style and simplicity for the benefit of very many, who are directly or indirectly connected with **NDT/NDE** methods/techniques.

I am one of those passionate admirers of the authors both Dr. Baldev Raj and Shri B. Venkataraman for their admirable professional abilities, authority on the subject and for the style of presentation with illustrative photographs/sketches. It is worth noting that the book has been very thoughtfully structured and interestingly almost every section therein contains crisply the summary of what you are going to read. At the end of each section, a set of good objective type questions has also been provided.

On behalf of every one in NCB including myself, I would like to place on record our sincere thanks and congratulate both the authors – Dr. Baldev Raj and Shri B. Venkataraman and also the others who have rendered help and support in bringing out a very useful and informative guide book on "Practical Radiography". I am particularly gratified because this is the first in **ISNT-NCB** series of books that follow.

I am certain that this book will be very useful not only to the students but also to all professionals in the NDT community and occupy a prominent place in all professional libraries.

K. Balaramamoorthy
Chairman, NCB

Preface

The wonderful discovery of X-rays by Sir Wilhelm Conrad Roentgen has revolutionised the field of NDT. Radiography today is a valuable diagnostic tool both in the field of medicine and industry. While in medical field, it is applied extensively for the detection of abnormalities in human body, in the industrial sector, it ensures the safety and reliability of the welds, components and plants. The aircraft and the trains in which we travel, the plants we operate are made safe through the employment of appropriate radiographic techniques. This technique is being applied successfully to microminiature parts like integrated circuit chips to mammoth missile components.

The reliability of any NDT technique depends on the judicious understanding and proper application of the technique. Training and certification is thus an essential part of ensuring quality through NDT. The first and foremost requirement of any training program is the availability of appropriate books and reference materials. The amount of knowledge gained by the NDT professional depends on the contents of the course notes and its style of presentation. Course notes are also essential for the harmonisation of training program nationally and internationally. Thus the course notes should reflect authenticity of the facts and also present the latest developments during that period in a lucid way.

Keeping in mind all the above mentioned facts, this book titled "Practical Radiography" has been written. It is intended as a practical guide for the beginner in the field – engineering or science students, technicians or any other NDT professional involved in radiography. Every effort has been made to present the matter in a clear and simple format with a large number of illustrations and examples. A novel feature is the summary of the important points on almost every page. Sample questions have been provided at the end of each Chapter. We are confident that this book would provide a valuable insight into the fundamental and practical aspects of radiography for the students and professionals. For those desirous of enhancing their knowledge further, a detailed reference list has been provided.

The contents of this book cover the syllabus that has been prescribed by the various international and national bodies under their training and certification schemes such as IS 13805, ISO 9712 and ASNT SNT-TC-1A. The book would thus serve as an essential basis for the harmonisation of training and certification program nationally and internationally.

A Request

Finally, a request for assistance from you; the valued reader. Our primary objective has been to find ways of presenting the central ideas of radiographic/ultrasonic technique in clear, interesting and effective fashion. Please do let us have your suggestions on additions, or deletions or corrections or simply put any aspect which would make the book more useful. Thus, when the second edition appears, you could be an important part of it. Our thanks to you in advance.

Baldev Raj

Acknowledgements

This is the first handbook on radiography published solely by the National Certification Board (NCB). The compilation and edits have been exclusively done by the authors and the Secretariat of NCB with the intent of reducing costs and controlling deadlines. In less than two years the ten sections comprising more than 20 individual Chapters were prepared. The result is now what you hold, a source of pride for the NCB. While it is the duty of NCB to publish such notes, the complete preparation is a herculean task. The real credit for this goes to the motivators namely Shri K.Balaramamoorthy, Chairman, NCB and Shri V.A.Chandramouli, Controller of Exams, NCB. The authors are grateful to Shri V.A.Chandramouli who had painstakingly gone through the entire draft. Many of the results published in this book are the works carried out at Indira Gandhi Centre for Atomic Research. The authors wish to place on record the unstinting help and support received from *S/Shri. S.Bagavathiappan, V.Chandrasekhar, N.Raghu, P.Balaraman, P.Sukumar*, of the Division for PIE & NDT Development, IGCAR, Kalpakkam and Shri *T.Rangachari* of HAL, Bangalore. Authors are thankful to Shri Dilip Takbate, M/s Pulse Echo Systems, India and to Dr. Brian F.Larson and the Centre for NDE, Iowa State University for their permission to reproduce photographs. Authors also thank all the members of the National Certification Board for their valuable suggestions.

Contents

SECTION 1
INTRODUCTION

After reading this section you should be able to answer

- ★ *Why NDT is needed ?*
- ★ *Advantages of using NDT,*
- ★ *Principles, applications and limitations of the CORE (6) NDT methods,*
- ★ *What are the various manufacturing and fabrication process,*
- ★ *Common defects arising during these processes.*

Chapter 1.1
Basics of Non-Destructive Testing

1.1.1 Need for Non Destructive Testing

It is well realised that all materials have defects and one of the principal uncertainties is the performance of the material/component in the presence of defects. Defects can be present in the raw materials itself or can be introduced during the various stages of processing and fabrication. Defects in materials can arise due to:

1. presence of discontinuities in the raw materials
2. improper manufacturing processes and component assembly
3. environmental and loading conditions during transport, storage and service

while some defects may be acceptable, some may not be acceptable. it is quite possible that a few of the unacceptable defects if left undetected can prove to be catastrophic. History is replete with various accidents in automobile and transportation industries and explosions of boiler/pressure vessels due to the presence of such defects and their propagation during service.

To get a product which is free from unacceptable defects, a means is thus necessary to monitor the quality of the product at every stage of its manufacture and during its service. This monitoring should be done in such a way that it does not impair the functionality or service performance of the product. Non-Destructive Testing (NDT) or Non-destructive Inspection (NDE) as the names imply, is the technology of assessing the soundness and acceptability of a material, component or structure without impairing its functional properties or 'Worth'. The term "NDT" includes many methods that can:

- detect surface or subsurface imperfections
- determine structure, composition, or material properties
- measure geometric characteristics

Akin to the medical field, where NDT is used as a diagnostic tool right from the inception of a child (ultrasonogram of child in the womb) to old age, NDT today is an inseparable part in the industrial world being used right from the cradle to tomb of a component i.e. right from the raw material stage through fabrication, during service and also to determine the remaining life of the component.

NDT is being practised for many decades. In the early days, the primary purpose of NDT was detection of defects. Critical

> Defect inherently present in material. Can arise due to
>
> * *presence of discontinuities in raw materials,*
> * *improper procedures during manufacture*
> * *environmental and service conditions*
>
> Non-Destructive Testing (NDT) or Non-Destructive Inspection (NDI) refers to testing a component without impairing its worth or Functional Properties.
>
> NDT - Needed to ensure the safety, reliability and quality of a component.

parts were produced with "safe life" design. The detection of defects during service was automatically a cause for removal of the component from service. The advent of fracture mechanics concepts in 1970's changed this scenario. A new design philosophy called "damage tolerant designs" came into existence. In this, components having known defects could continue to be used as long as it could be established that these defects would not grow to a critical size that would result in catastrophic failure. It thus became possible to accept structures with defects provided the defect size could be monitored to ensure that it is well within the safe limits. A new demand was thus placed on the NDT community. Mere detection of flaws was not enough. It was necessary to characterise the defect and obtain quantitative information about its size, shape and location so that this could serve as an input to fracture mechanics calculations for predicting the remaining life of the component. This need was particularly strong in the nuclear, defence and space industries. This led to the emergence of Non Destructive Evaluation (NDE) as a new discipline. NDE refers to quantitative inspection in which the defect is not only detected, but also characterised with respect to its size, shape and orientation. Today, a number of research programs have evolved all over the world in the field of NDE.

1.1.2 Benefits of NDT

The main advantage of applying NDT for material inspection, during manufacturing process and for inservice inspections is that it :

1. ensures product quality, reliability and safety
2. aids in optimum product design.
3. controls manufacturing processes to within specified tolerances
4. ensures uniform quality levels
5. ensures customer satisfaction
6. predicts impending failures, thus preventing costly shutdowns and aids in plant life extension.

A wide range of industries/professions use NDT methods. To name a few: Nuclear, Aerospace, Automotive, Chemical, Defence, Electronics, Electrical, Fabrication, Fertilisers, Food Processing, Marine, Medical, Metals & Non-Metals, Petrochemical, Power, Security, Surface Transport. Indeed, progress of human civilisation in many industrial activities would not have been possible without diagnostic NDT. The success of NDT depends on the combination of **QUALIFIED PERSONNEL**, Calibrated Equipment, Documented practices and Standard Procedures and Codified Acceptance Criteria.

Non Destructive Evaluation (NDE) refers to quantitative NDT in which the defect is characterised with respect to its size, shape and location.

Success of NDT possible only by a combination of

* *Qualified personnel*
* *Calibrated equipment*
* *Documented practices and standard procedures*
* *Codified Acceptance criteria*

Most widely used NDT methods are :

* *Visual Examination,*
* *Liquid Penetrant,*
* *Magnetic Particle,*
* *Radiography,*
* *Ultrasonics and*
* *Eddy Current Testing.*

1.1.3 NDT Methods

There are several established NDT methods currently used by the industry. With the development of newer materials and increasing demands on quality due to reduced margins on safety and stringent specifications, newer methods are being constantly researched and developed. Many of the established methods have had their effectiveness and efficiency enhanced by better sensor technologies, improved electronics and instrumentation, computerisation and robotic attachments. While there are more than 40 NDT techniques, forming the core of NDT are six basic methods: Visual testing (VT), Liquid Penetrant Testing (LPT), Magnetic Particle Testing (MPT), Radiographic Testing (RT), Ultrasonic Testing (UT) and Eddy Current Testing (ECT). In this Chapter, we briefly outline the principles, advantage and limitation of these widely used methods.

1.1.3.1 Visual Examination (VE)

Visual examination – the first stage in the examination of a component is one of the simplest methods used for the inspection of defects / discontinuities open to the surface. The flaws present on the surface are more detrimental compared to the flaws located entirely below the surface because:

1. surface is the part where maximum stress is present,
2. surface is the part that is connected to environment. Hostile environment may introduce corrosion, erosion, etc. and
3. many studies have revealed that small imperfection on the surface has served as nucleus for fatigue crack initiation and such cracks have lead to catastrophic failure.

The method finds wide applications for inspection of materials during various stages of manufacture. For many non-critical parts, integrity is verified principally by visual inspection. Even when other NDT methods are used, visual inspection still constitutes an important part of practical quality control. VE can and should be done before, during and after manufacturing. VE is useful for assessing the following.

(a) dimensional accuracy of the components
(b) conformity of components to size, and contour requirements
(c) acceptability of part appearance with regard to surface roughness and finish
(d) detection of surface breaking cracks and other defects in castings, forgings and weldments.
(e) determining the weld profiles externally and internally

A variety of aids are available to enhance the capabilities of visual inspection. These range from simple aids such as magnifying glass, optical microscopes to fiberoscopes, videoimagescopes and robotic based remote inspection systems for closely inspecting the internal of tubes, pipes and areas with limited access.

Surface lighting and angle of inspection are two of the important factors to be considered during visual inspection. The surface should be inspected at an angle which is preferably at 90 degrees to the surface of inspection. Adequate lighting should be

> Visual Testing first stage in examination of component.
>
> Visual aids range from simple magnifiers, microscopes to sophisticated borescopes, fiberoscopes to Remote vision systems.
>
> Surface lighting, angle of inspection and the vision acuity of the personnel are three of the critical parameters in visual inspection.

available on the inspection surface (greater than 200 lux). For critical inspections, more than 500 lux should be available on the inspection surface. Visual acuity of the personnel plays a very important role in the efficiency of defect detection. Personnel involved in visual examination should undergo vision acuity tests at least once in a year.

1.1.3.2 Liquid Penetrant Testing (LPT)

LPT is one of the simplest, but highly sensitive NDT methods used for the detection of defects open to the surface. The method is applicable to almost any component, large or small, of simple or complex configuration, and it is employed for the inspection of wrought and cast products in both ferrous and non-ferrous metals and alloys, ceramics, glassware and some polymer components.

Principle

In this method, the surface of the component is cleaned well and a visible or fluorescent dye solution called Penetrant is applied. The penetrant is allowed to remain on the surface for a certain period of time referred to as the dwell time. During this time, the penetrant seeps into the discontinuities by capillary action. After the dwell time has elapsed, the excess penetrant is removed from the surface. A developer (dry or wet) is applied uniformly over the surface. The developer acts as a blotter, drawing the trapped penetrant out of the discontinuity on to the surface. This is referred to as bleedout. With visible dyes, colour contrast (usually red) between the penetrant and the developer makes the discontinuity to be seen under natural or artificial lighting. With fluorescent dyes, ultraviolet light is used to make the bleedout fluoresce brightly, thus allowing imperfections to be seen.

The various steps in penetrant processing are schematically indicated in Fig. 1.1.1 below. Penetrants are classified as visible or fluorescent depending on the nature of the dyes used. While visible penetrants use a colour contrast (usually red) dye, fluorescent penetrants use a dye which fluoresces under black light. Fluorescent penetrants have higher sensitivity compared to visible dye penetrants. A number of penetrant types have been developed to cater for the wide variety of inspection conditions that can occur in practice. The main types are water-washable, post-emulsified and solvent based systems. The above systems differ from each other depending upon the method by which excess penetrant is removed. In water washable system, the penetrant is removed by a rinse of water. The post-emulsified penetrant system requires a separate emulsification step, which makes the penetrant soluble in water. In solvent removable system, a solvent is used to remove the excess penetrant. In terms of sensitivity, the water washable penetrant system is the least sensitive while post emuslfied systems have the highest sensitivity.

> Penetrant Testing based on the principle of capillary action.
>
> Suitable only for detection of defects open to the surface.
>
> Two types of Penetrants *Visible and Fluorescent*
>
> Fluorescent penetrant more sensitive
>
> Penetrant systems classified as water washable, post emulsifiable and solvent removable.
>
> Post emulsified is the most sensitive and water washable system least sensitive.

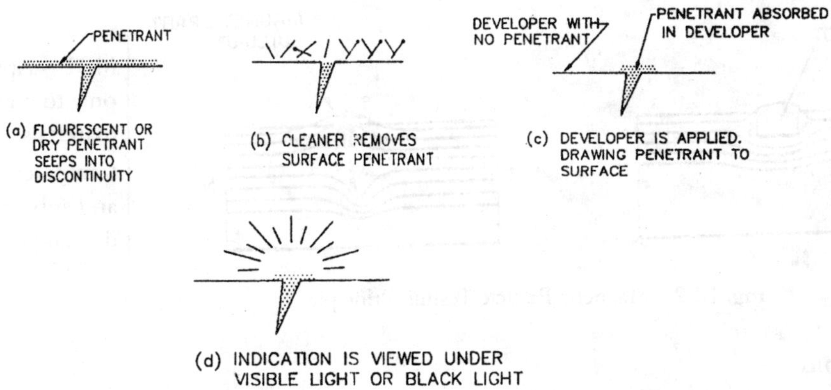

Fig. 1.1.1 Schematic sketch of LPT methodology

Applications

a. LPT is well suited for detection of all types of cracks, laps, porosity, shrinkage areas, laminations and other similar defects open to the surface.

b. Can be applied on any material regardless of size, shape, thickness and composition.

c. It is highly sensitive and very cost effective

Limitations

a. It cannot detect subsurface defects and

b. It is difficult to apply on rough and porous surfaces.

1.1.3.3 Magnetic Particle Testing (MPT)

MPT is a sensitive test method used for the detection of surface and sub-surface defects in ferromagnetic materials and components. It is based on the principle that when a ferromagnetic material is placed in a magnetic field, it gets magnetised. The lines of force through the material is continuous from one pole to another. When any surface or subsurface defect is present, the magnetic field and the associated lines of force get deflected and forms a leakage field. This leakage field is present at and above the surface of the magnetized component. Its presence can be visibly detected by the utilization of finely divided magnetic particles. Dusting the surface with iron particles either dry or suspended in liquid results in the collection of the magnetic particles at the discontinuity. The 'magnetic indication' so formed can be clearly seen and reveals the location, size and shape of the discontinuity. Magnetisation of a component is accomplished using alternating current (AC), direct current (DC), Half Wave DC (HWDC), permanent magnet or electromagnets.

Applications

MPT is very effective for the detection of surface and subsurface cracks, grinding cracks, heat treatment cracks, stringer type non-metallic inclusions, porosities, laps and folds. The method can be applied to finished articles, billets, hot rolled bars, castings and forgings.

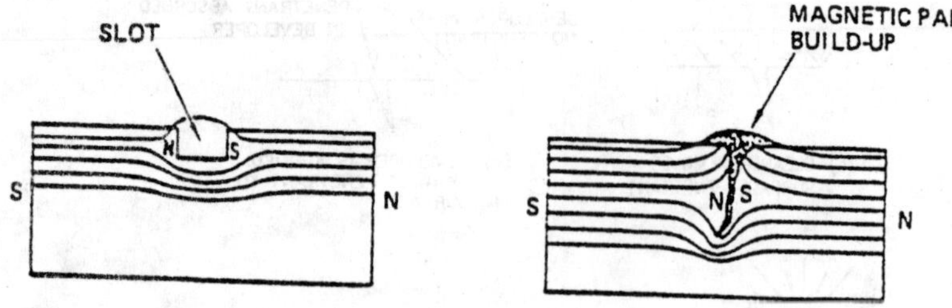

MPT – Applicable only to magnetic materials.
Can detect surface and sub-surface defects

Fig. 1.1.2 Magnetic Particle Testing Principle

Limitations

a. It can be used only for ferromagnetic materials.
b. Magnetic fields must be perpendicular to the principal plane of discontinuity. Two or more sequential inspections is required to cover all orientations of defects.
c. Demagnetization and post cleaning following inspection is often necessary
d. Exceedingly large currents are required for very large parts. Thus, care is necessary to avoid local heating and burning of finished parts or surface at the points of electrical contact.

1.1.3.4 Radiographic Testing (RT)

Radiography is one of the oldest and most widely used NDT methods. In this method, a source of penetrating radiation such as X-rays or gamma rays is used. When these radiations pass through an object, they are differentially absorbed. The amount of radiation absorbed depends on the nature of the object, its composition (density) and thickness. Variations in the beam intensity are detected by a detector, normally a sheet of photographic film. On processing the film, a permanent image is formed. The developed film is called the radiograph. Possible imperfections are indicated on the film as density changes in the same manner as a medical X-ray film shows broken bones or stones in kidney. Thus during material testing, a void would appear darker since more radiation will pass through while a tungsten inclusion because it is of higher density will appear to be white since lesser radiation will pass through. Fig. 1.1.3 shows the schematic setup for radiographic inspection. The commonly used sources are X-rays or gamma rays. X-rays are produced when fast moving electrons are suddenly brought to rest in X-ray machines. Gamma rays are produced during decay of radioactive isotopes such as Iridium 192 and Cobalt 60.

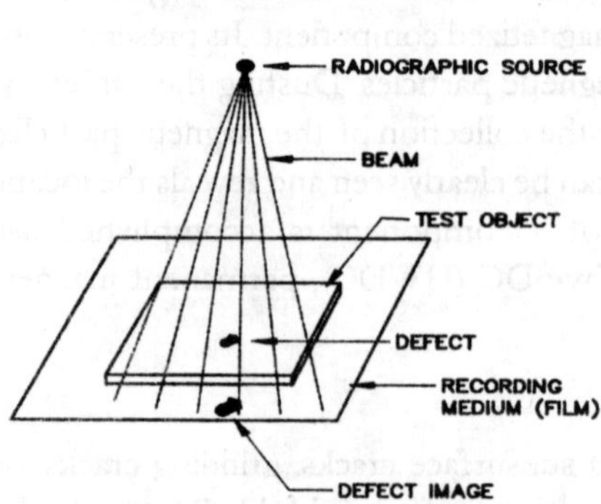

Fig. 1.1.3 Schematic setup for radiographic inspection

Applications

a. Radiography is widely used for the inspection of welds, castings, assemblies and adhesively bonded structures, plastics, wood, composite materials and electronic components.

b. Objects can vary in size from micro miniature parts such as IC chips to huge components such as rocket motor casings.
c. It is the best technique for the detection of volumetric defects such as voids and porosities.
d. A visual image is formed which makes interpretation easy and provides a permanent record.

Radiography based on the principle of differential absorption of penetrating radiation.

Best suited for detection of volumetric defects such as voids.

Provides a visual image and permanent record

Widely used in weld and casting inspection.

Linear/planar defects are difficult to detect.

Radiation hazards pose safety problems.

Limitations

a. Radiography requires access from both sides of the component.
b. Linear flaws are difficult to detect; for e.g., cracks cannot be detected unless they are oriented inline to the radiation beam.
c. Compared with other (NDT) methods of inspection, radiography is expensive. Since radiations are hazardous, proper precautions are to be taken while doing radiography.

1.1.3.5 Ultrasonic Testing (UT)

In this inspection method, high frequency sound waves are sent into the object under test. The sound waves travel through the material. During their path of travel they suffer loss of energy and are reflected at interfaces. A receiver probe picks up the reflected wave and an analysis of this signal is done to locate flaws in the object under inspection. Sound waves follow the laws of optics in their propagation. Further, the velocity of propagation of sound in various metals has been very accurately determined. The time taken by a sound pulse to travel through a material is a direct measure of the length of path travelled by it. In ultrasonic inspection both through transmission and pulse-echo techniques are used.

Most of the ultrasonic inspection is done in the mega cycle range at frequencies between 0.5 and 25 MHz which is well above the audible range. The probes used for ultrasonic inspection contains a piezoelectric crystal. The transmitter probe generates ultrasonic pulses and transmits them into the materials while the receiver probe receives the reflected pulses.

Applications

Ultrasonics can detect defects such as cracks, laminations, shrinkage, cavities, flakes, pores and other discontinuities, inclusions, etc. in plates, pipes, welds, castings and forgings reliably. The technique is widely used for quality control and material inspection in all major industries. It is used very extensively for in-service inspection in the nuclear and process industries. The availability of compact battery operated equipments has made this a versatile tool for field applications. On-line techniques have also been developed to monitor and classify materials as "acceptable" or "reject" and for process control applications.

Limitations

(a) Manual operation requires careful attention by experienced technicians.

(b) Unfavorable geometries, coarse anisotropic structures, parts that are rough or irregular in shape or very small or thin are difficult to examine.

(c) Couplants are needed to provide efficient transfer of ultrasonic wave energy between transducers and parts being inspected.

(d) Reference standards are needed for calibrating the equipment and for characterising the flaws.

1.1.3.6 Eddy Current Testing (ECT)

> Ultrasonic Testing based on the reflection of sound waves.
>
> Widely used for detection of defects in castings, forgings, pipes, plate and welded components.
>
> Difficult to use on coarse, rough, anisotropic structures or small or thin parts.
>
> Needs couplant for efficient ultrasonic energy transfer and reference standards for calibration.

This technique is based on induction of currents in electrically conducting material being inspected and observing the interaction between these currents and the materials. When magnetic flux through a conductor changes, induced currents are setup in closed paths on the surface of the conductor. These currents are in a direction perpendicular to the magnetic flux and are called as Eddy Currents. When an alternating current is passed through a coil, a magnetic field is setup around it. The direction of magnetic field varies with each cycle of the alternating current. This is known as the excitation field. When a conductor is brought near this field,, eddy currents are introduced in it. The direction of the eddy current changes with the direction of magnetic field during each cycle of AC. These eddy currents produce their own magnetic field, which oppose the excitation field. Thus the resultant field is changed which changes the coil impedance. This change in impedance is suitably displayed using a meter or a CRT screen.

The test coil also referred to as the probe is the main link between the test instrument and test object and serves two main functions- the first to establish a varying electromagnetic field which induces eddy currents within the test object and the second is to feed the response due to the electromagnetic field to a signal analysis system. Various types of probes are available suitable for the inspection of flat surfaces and tubular products. ECT is used extensively to identify or differentiate between a wide variety of physical, structural and metallurgical conditions in conductive materials.

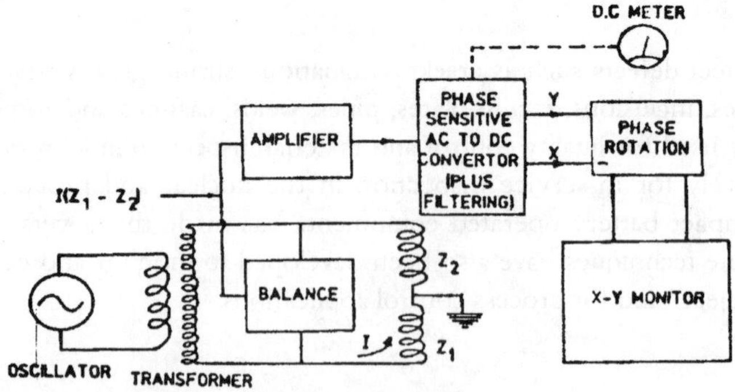

Fig. 1.1.4 Block diagram of ECT instrument

Applications

Eddy current testing is used

(a) to measure or identify such properties as electrical conductivity, magnetic permeability, grain size, heat treatment conditions, hardness and physical dimensions;

(b) to detect seam, laps, cracks, voids and inclusions and other surface discontinuities in non ferromagnetic material

(c) to sort dissimilar metals, detect differences in composition, microstructure and other properties

(d) to measure the thickness of a non-conductive coating on a conductive metal

Limitations

The main limitation is its low penetration in parts being examined being limited to thin walls or near surface flaws. The technique is difficult to use on ferromagnetic materials. False indications are possible because of mixed variables, edge effects and lift off effects. Extensive technical knowledge is required for the development of inspection techniques; design of specific probes and to interpret the inspection data.

ECT based on the induction of currents in electrically conducting material.

Widely used for the inspection of tubes for defects and wall thinning, sorting of materials and measurement of coating thickness

Limited to detection of near surface flaw and is affected by edge effects and lift off effects.

Difficult to apply in case of ferromagnetic materials

Chapter 1.2

Basic Metallurgical Processes and Defects

The purpose of NDT is to ensure the worthiness of a component without impairing its functionality. Once a indication has been detected by NDT, it is to be judged whether the indication is relevant, non relevant or false. In case the indication is relevant, then it is to be evaluated as per the acceptance criteria prescribed by relevant codes or standards. Based on this evaluation, the component is accepted or rejected. Although it is not generally considered to be a part of non destructive testing, a reasonable working knowledge of methods of fabrication is necessary for the NDT personnel to appreciate the types and likely locations of faults when a particular form of fabrication is used and for proper evaluation. This chapter briefly outlines the primary metallurgical processes and the defects that are likely to arise during these processes. A few terminologies that are commonly used in connection with the indications are also given.

1.2.1 Definitions

Discontinuity is any local variation in material continuity such as change in geometry, structure, composition or properties or presence of holes, cavities or cracks. Discontinuities are not always bad or hazardous and may even sometimes be needed in the design or may be helpful in some kinds of processing.

Defect – When any discontinuity, single or multiple, is of such size, shape, type and location that it can create material failure during service, it is commonly called a "DEFECT". Finding defects is one of the primary objectives of NDT. It must be understood, however, that a fault that is a defect under one set of conditions of service may be only a simple discontinuity that is not harmful in a different application.

False indication – an indication detected by an NDT method that is interpreted to be caused by a condition other than a discontinuity

Flaw – Is a imperfection or discontinuity that can be detected by NDT. A flaw detected by NDT need not be necessarily rejectable.

Imperfection – it is a departure of a quality characteristic from its intended or designed condition

Non relevant indication – an indication because of an imperfection or discontinuity that has been detected by an NDT method and which is not rejectable.

Relevant indication – an indication because of an imperfection or discontinuity that has been detected by an NDT method and which requires further evaluation.

1.2.2 Metallurgical Processes and Defects

1.2.2.1 Casting

A casting may be defined as a piece of metal shaped in a mould. The mould is a patterned or shaped cavity which receives the molten metal. The process of casting involves pouring or injecting the molten metal into the mould and allowing it to solidify. The cavity or mold may be of intricate shape so that when the solidified metal is removed a part is produced which, with or without further preparation, may be used for its intended purpose. The basic steps involved in the casting process are:

a. preparation of a pattern or die to form the final shape and size of part required.
b. preparation of the mould cavity using the pattern and suitable mould materials.
c. heating and melting of the metal.
d. pouring/injecting the molten metal into the mould cavity.
e. removal and, when necessary, the further cleaning or shaping of the part.

Casting processes can be grouped into two categories (a) those using a non-permanent mould which is used only once, and (b) those using permanent moulds in which the mould can be used repeatedly to obtain a number of casting pieces. Examples of non-permanent mold castings include:

Sand casting
Shell mould casting and
Investment or precision casting

Permanent mould castings include:

Centrifugal castings
Die or pressure castings

The flaws, which may be formed during the casting processes include (a) Inclusions (b) Porosities (c) Shrinkage flaws (d) Cold shut (e) Hot tear (f) Cracks (g) Blowholes (h) Segregations and (i) Mottling.

1.2.2.2 Forging

Forging is the working of metal into a useful shape by hammering or pressing. It is the oldest metal forming process. Most forging operations are carried out hot, although certain metals can be cold forged. The hot working of metals in the forging process results in an improvement in the mechanical properties. This method of shaping is therefore used in the manufacture of parts requiring good mechanical properties. Improvement in the mechanical properties results form a general consolidation of the metal and closing of gas and contraction cavities by means of mechanical pressure, a refinement of the crystal structure, a destruction of the continuity of intergranular concentrations of impurities and inclusions.

Forging is done using two major classes of materials - the forging hammer or a press. The forging hammer or drop hammer, delivers rapid impact blows to the surface of the metal, while the forging press subjects the metal to a slow-speed compressive force. Forging may be considered under two categories. First where the working surface of the dies is flat or of uniform curved contour and shaping is done by manipulation using tools of simple shape. This is called 'open-die' forging. The second is where impression dies are used and the metal is shaped by being forced into the die impressions. This is called closed-die forging. Closed-die forging can produce precision forging with close dimensional tolerances. The defects that can occur during forging are (a) forging lap and (b) forging bursts or cracks

1.2.2.3 Extrusion

Extrusion is the process by which a block of metal is reduced in cross section by forcing it to flow through a die orifice under high pressure. In general, extrusion is used to produce cylindrical bars or hollow tubes. However, shapes of irregular cross section may be produced from the more readily extrudable metals like aluminium. Because of the large forces required in extrusion, most metals are extruded in hot conditions where deformation resistance of the metal is low. However, cold extrusion is possible for many metals and has become an important commercial process. Defects encountered during extrusion include internal piping or extrusion defect and cracks.

1.2.2.4 Rolling

The process of plastically deforming metal by passing it between rollers is known as "Rolling". Since the metal is formed by a squeezing action, rolling can be considered as a continuous forging process with the rolls acting as hammers and the metal being drawn down. Rolling may be performed above the temperature of recrystallisation (hot rolling) or below the temperature of re-crystallisation (cold rolling). Hot rolling is always used for the initial rolling of the cast ingot. Not only is it easier to break down the ingot to size quickly when it is hot and plastic, but the hot-rolling process closes any casting discontinuities and forge welds their surfaces together. This prevents any faults, which could lead to lamination, being carried forward into subsequent rolling operations. The cold rolling of metals has reached a position of major importance in industry. By cold rolling, sheets, strip and foil with good surface finish, increased mechanical strength and precise dimensions are being produced. Some of the commonly occurring defects during the rolling process includes laminations, stringers and seams.

1.2.2.5 Drawing

Drawing operations involve pulling a metal through a die by means of a tensile force applied on the exit side of the die. Most of the plastic flow is caused by compressive force, which arises from the reaction of the metal with the die. The reduction in diameter of a solid bar or tube by successive drawing is known as "bar or wire or tube drawing". Two of the typical drawing defects are seams and centre-burst or "chevron cracking" also called "cupping".

1.2.2.6 Welding

This is one of the oldest and most widely used materials joining process. The American Welding Society definition for a welding process is "a materials joining process which produces coalescence of materials by heating them to suitable temperatures with or without the application of pressure or by the application of pressure alone and with or without the use of filler material". Generally same materials are joined together by welding. In some special cases dis-similar welding is involved to get required properties (Ex: Carbon steel to alloy steel/stainless steel). Also, in other cases certain materials are overlayed on a base material (Ex: Stainless steel overlay on carbon steel, stellite overlay on stainless steel).

Welding can be classified as either "manual", "semi-automated" and "automated". These terms refer to the method of control and guidance. In manual welding, the entire welding operation is performed and controlled by hand. In semi-automated welding, the welding operation is manually controlled but the equipment controls the filler metal feed. In automated welding, the weld is made with equipment that controls and performs the entire welding operation without constant observation and adjustment of the controls by an operator. There are various types of welding processes such as Shielded Metal Arc Welding (SMAW), Gas Tungsten Arc Welding (GTAW), Gas Metal Arc Welding (GMAW), Submerged Arc Welding (SAW) and Electro-Slag Welding. Some of the newer processes include Electron Beam Welding (EB Welding), laser welding, friction welding and plasma arc welding.

There are five basic types of joints. These are the butt joint, edge joint, fillet or tee joint, corner joint and the lap joint. A number of variations of these are possible depending on the manner of preparation and assembly.

The general sources of weld defects include: improper design, poor joint preparation, defects in parent material, improper welding technique, faulty solidification of molten metal, and heating and cooling effects on both the base metal and weld metal. The possible defects that can occur in fusion weldments are (a) Crater cracks (b) Stress cracks (c) Porosity (d) Slag inclusions (e) Tungsten inclusions (f) Lack of fusion (g) Lack of penetration and (h) Undercut.

It is clear from above that defects are likely in any metallurgical process. Table 1.1.1 provides a summary of the typical defects occurring in various metallurgical processes. By the judicious use of non destructive testing methods and techniques, it has been possible to decrease the factor of

Table 1.1.1 Summary of typical defects introduced during various processes

PROCESS	DEFECTS POSSIBLE
Manufacture of raw materials	Piping, Cracks, Shrinkage, Porosity, Inclusions, Segregation, Metallurgical variations
Manufacture of components.	Machining Faults, Heat treatment defects, Welding defects, Residual stress cracks
Component assembly.	Missing parts, Incorrectly assembled parts Additional welding defects, Additional stress cracking
Sheet and plate fabrications	Thickness/density/compositional variations, Holes, Inclusions, Laminations, Seams, Cracks, Pit Marks
Fabrication of Bars and tubes	Pipe, Cupping, Inclusions, Cracks
Casting	Cold shuts, Surface cracks, Shrinkage Cavities, Porosity, Core shift, Hot tears, Inclusions
Forging	Laps, Inclusions, Internal bursts, Internal flakes, Cracks and Tears
Welding	Cracks, Inclusions, Lack of Fusion, Porosity, Lack of penetration, Burn Through, Undercuts, Crater Crack
Service	Fatigue cracks, Stress corrosion cracks, Blistering, Thinning, Corrosion Pits, wear, creep

ignorance about material soundness without decreasing the factor of safety in the finished product. With QUALITY becoming the password to global success, the use of NDT techniques has been recognized by management as a means of meeting consumer demands for better products, reduce cost and increase production. Apart from ensuring quality, NDT is also being used to aid in new process development and manufacturing techniques.

Chapter 1.3

Materials in Service

Materials are required to operate and perform in widely varying environmental conditions. The requirements of safety and reliability demand that the materials and components should perform well in their environments and situations without premature failure. There are a number of factors and processes which can cause the failure of materials. Apart from loss of production and wastage of manhours, premature failures of critical components can also prove to be catastrophic resulting in accidents and loss of lives. It is thus essential to understand the cause of failure and try to control the same.

1.3.1 Conditions Leading to Defects and Failure

The rapid advances in the field of NDE sensors and instrumentation coupled with the greater understanding of materials science, metallurgy and engineering has led to reduced percentage of failures. However, failures still continue to occur. Failures can be classified into three categories:

(a) Operational failures – caused by overload, wear corrosion and stress corrosion, brittle fracture and metal fatigue.

(b) Failures due to improper design. Factors to be considered in this include presence of sharp corners, high stress areas, safety stress factor and choice of material.

(c) The third type of failure is caused by thermal treatments such as forging, hardening, tempering and welding, and by surface cracks caused by grinding or other means. These aspects and especially those related to operational or in-service conditions are briefly outlined below.

1.3.2 Corrosion

With the exception of some noble metals, all metals are subject to the deterioration caused by ordinary corrosion. Iron, for example, tends to revert back to its natural state of iron oxide. Other metals revert to sulfides and oxides or carbonates. Buildings, ships, machines and automobiles are all subject to attack by the environment. The corrosion that results often renders them useless and they have to be scrapped. Crores of Rupees a year are lost as a result of corrosion. Corrosion can also lead to dangerous situations, such as on bridges, where the supporting structures have been eaten away, or in aircraft in intergranular corrosion can weaken the structural members of the aircraft which can result in sudden failures.

1.3.3 Fatigue

When metal parts are subjected to repeated loading and unloading over prolonged periods they may fail at stresses far below their yield strength with no sign of plastic deformation. This is called as fatigue failure. When designing machine parts that are subject to vibration or cyclic loads, fatigue strength is critical than ultimate tensile or yield strength. Fatigue is a universal phenomenon observed in most solids.

Failure caused by fatigue are found in many of the materials of industry. Fatigue failure is primarily caused by a crack that is initiated at a notch, bend, scratch, or any defect with a sharp edge that continues to grow gradually as a result of stress reversals on the part. The crack growth continues until the cross-sectional area of part is reduced sufficiently to weaken the part to the point of failure. Fatigue is greatly influenced by the kind of material, grain structure and service conditions.

1.3.4 Wear

Wear may be defined as undesired removal of material from contacting surfaces by mechanical action. Excessive wear can be caused by continuous overload, but wear is ordinarily a slow process that is related to the friction between two surfaces. Rapid wear can often be attributed to lack of lubrication or the improper selection of material for the wear surface. Some wear is to be expected, however, and could be called normal wear. Wear is one of the most frequent causes of failure.

Different types of wear include abrasive wear, erosive wear, corrosive wear and surface fatigue. Various methods are used to limit the amount of wear in the part. One of the most commonly used methods to reduce wear is simply to harden the part. Other methods of limiting wear are electroplating (especially the use of hard industrial chromium) and anodizing.

1.3.5 Overload

Overload failures are usually attributed to faulty design, extra loads applied, or an unforeseen machine movement. Shock loads or loads applied above the design limit are quite often the cause of the breakdown of machinery. Although mechanical engineers always plan for a high safety factor in designs (for instance, the 10 to 1 safety factor above the yield strength that is sometimes used in fasteners), the operators of machinery often tend to use machines above their design limit. Of course, this kind of overstress is due to operator error. Inadequate design can sometimes play a part in overload failures. Improper material selection in the design of the part or improper heat treatment can cause some failures when overload is a factor. Often a machinist or welder will select a metal bar or piece for a job based upon its ultimate tensile strength rather than upon its yield point. In effect this is a design error and can ultimately result in breakdown.

1.3.6 Brittle and Ductile Fracture

Fracture preceded by a significant amount of plastic deformation is known as ductile fracture, otherwise it is brittle fracture. Brittle fracture occurs, when plastic flow is inhibited either by the effective locking of atomic dislocations by precipitates or elements or by the pre-existence or formation of cracks and imperfections acting as local stress raisers in the material. All materials can be embrittled if the temperature is lowered sufficiently. Glass, sealing wax, germanium, silicon and other materials though ductile at temperatures close to their melting point are brittle at ordinary temperatures.

Questions

Choose the most appropriate answers from the options given.

1. Short wavelength electromagnetic radiation produced during the disintegration of nuclei of radioactive substances is called:
   a. X- radiation b. gamma radiation.
   c. scatter radiation. d. back-scatter radiation. e. beta radiation.
2. Which of the following materials cannot be tested by magnetic particle techniques?
   a. High alloy steels b. Tool steels
   c. Copper alloys d. Ferritic stainless steels
   e. both c and d above.
3. Materials, which are strongly attracted to a magnet, are called:
   a. para magnetic materials. b. non-magnetic
   c. ferromagnetic d. diamagnetic
   e. all of the above
4. Magnetic particle inspection methods are recognized as superior to liquid penetrant techniques when the:
   a. surfaces of the test object are corroded b. surface is anodized.
   c. parts are painted. d. part is made from austenitic steel.
   e. parts are of complex shape
5. The tendency of a liquid penetrant to enter a discontinuity is primarily related to:
   a. the viscosity of the penetrant b. capillary forces.
   c. the chemical inertness of the penetrant. d. the specific gravity of the penetrant.
   e. all of the above.
6. Liquid penetrant testing is a non-destructive test, which can be used for:
   a. detecting all types of discontinuities in a test specimen.
   b. determining the length, width and depth of discontinuities in a test specimen.
   c. determining the defect severity
   d. locating discontinuities open to the surface
   e. all of the above
7. Which of the following is not a characteristic that applies to liquid penetrant testing?
   a. This method can accurately measure the depth of a crack or discontinuity.
   b. This method can be used for testing large parts in the field.
   c. This method is best suited for defects that are open to the surface.
   d. This method can be made more or less sensitive by using different penetrant materials.
   e. This is one of the simplest NDE methods
8. Eddy currents are circulating electrical currents induced in conductive materials by:
   a. continuous direct current. b. alternating current
   c. an alternating magnetic field. d. a piezoelectric force
   e. any of the above.

9. Eddy current testing relies on the principle of:
   a. magnetostriciton.
   b. electromagnetic induction.
   c. piezoelectric energy conversion
   d. magnetomotive force.
   e. direct current flow in a resistive circuit.
10. In eddy current testing, discontinuities can be reliably detected when the eddy currents are:
    a. coplanar with the major dimension of the discontinuity.
    b. perpendicular to the major plane of the discontinuity
    c. parallel to the major dimension of the discontinuity
    d. 90° out of phase with the current in the coil.
    e. none of the above
11. The most widely used method for the detection of linear/planar defects in welds is
    a. radiography
    b. ultrasonics
    c. eddy current testing
    d. magnetic particle testing
    e. penetrant testing
12. In penetrant inspection, the most sensitive fluorescent penetrant system is
    a. water washable
    b. post emulsifiable
    c. solvent removable
    d. none of the above
    e. all are equally sensitive
13. Most appropriate NDT technique is chosen based upon
    a. type of material
    b. geometry of component
    c. nature of defect expected
    d. accessibility for examination
    e. all of the above
14. The technique of industrial radiography uses
    a. electromagnetic radiations
    b. particulate radiography
    c. X- and gamma radiations
    d. UV and light radiations
    e. all of the above
15. The most simplest and least expensive NDT technique which essentially precedes all other inspections is
    a. liquid penetrant
    b. visual testing
    c. magnetic particle
    d. ultrasonic testing
    e. radiographic testing

SECTION 2

PHYSICAL PRINCIPLES OF THE TEST

After reading this section you should be able to answer

- ★ *What are the constituents of an atom*
- ★ *What is radioactivity*
- ★ *Explain the concepts of specific activity, half and tenth value layers*
- ★ *How radiations such as alpha, beta, X- and gamma interact with matter*
- ★ *Methods of detecting radiation*
- ★ *Principle of gas filled and scintillation detectors*

Chapter 2.1

Basic Radiation Physics

A discussion on the constituent parts of an atom is important in understanding the fundamental aspects of radiography.

2.1.1 Atom

An atom is defined as the smallest unit of matter that represents the physical and chemical properties of an element. The atom is composed of a central positively charged core known as the nucleus. Travelling in a path around the nucleus in well defined orbits or energy shells are one or more smaller, negatively charged particles called electrons. The nucleus consists of two types of fundamental particles - the positively charged 'protons' and the electrically neutral 'neutrons'. Protons and neutrons are called by the common name 'nucleons'.

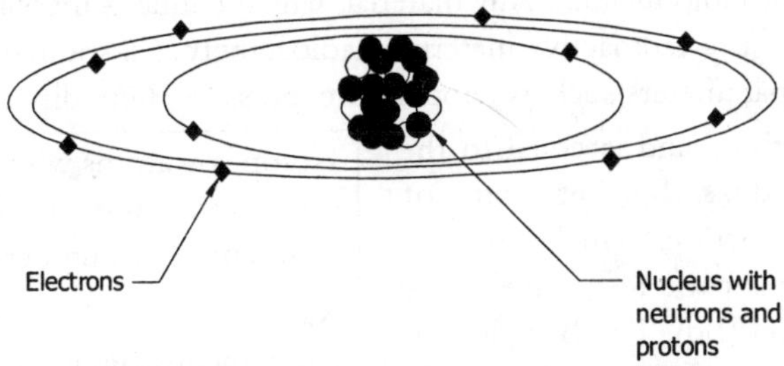

Fig. 2.1.1 Typical structure of an atom

> Atom – *smallest unit of matter that represents the physical and chemical properties of an element.*
>
> Consists of Nucleus comprising of positively charged Protons and Neutrons (no charge)
>
> Electrons in well defined orbit revolve around the nucleus
>
> Atom–*electrically neutral*
>
> The number of protons in an atom is the atomic number (Z)
>
> The mass number is the sum of atomic number and number of neutrons.

An atom is electrically neutral as the number of protons in its nucleus is equal to the number of electrons surrounding it. The structure of a typical atom is shown in Fig. 2.1.1. The number of protons in an atom is known as its atomic number (Z). Every atom of a particular element has a definite number of protons eg. Sodium has 11, Aluminum has 13, Cobalt has 27 and Iridium has 77.

The proton has a mass of 1.673×10^{-24} gms and carries a positive electrical charge of 1.6×10^{-19} coulombs. The neutron has a mass of 1.6×10^{-24} gms and no electrical charge. The electron has a mass of 9.107×10^{-28} gms about 1/1840 of the mass of proton and has a negative electrical charge of the same magnitude as that of proton, i,e. about 1.6×10^{-19} coulombs. Thus, the mass of an atom is almost wholly concentrated in the nucleus, as the electrons have negligible mass. Atom of any element X is represented as X_z^A where A is the mass number. The mass number of an atom is the sum of number of protons (Z) and the number of neutrons (N) in its nucleus i.e.

$$A = N + Z \qquad \text{... (2.1.1)}$$

2.1.2 Isotopes

The chemical behaviour of an element is dependent purely on the atomic number. Any atom or nuclide having the same number of protons, and thus the same atomic number Z, but different number of neutrons is called an isotope of the element. For example the three important isotopes of the element hydrogen whose atomic number is one are $_1H^1$ (hydrogen), $_1H^2$ (deuterium) and $_1H^3$ (tritium).

2.1.3 Radioactivity

Some isotopes are stable while some others are unstable. The stability of an isotope depends upon the ratio of neutrons to protons in its nucleus. An unstable isotope becomes stable by emitting various types of ionising radiations such as alpha, beta, X- or gamma radiations. This process of decay is known as radioactivity. Isotopes, which undergo radioactive decay are called radioactive isotopes or radioisotopes or radionuclides. Any material, which contains measurable amounts of one or more radionuclides is a radioactive material. Radioactivity is a spontaneous process. It is not affected by external parameters such as temperature, pressure, humidity etc.

Curie (Ci) is the unit of radioactivity and is equal to the disintegration rate of 3.7×10^{10} d.p.s. Smaller units of radioactivity are millicurie and microcurie, which are one Thousandth and one millionth part of a curie respectively. The SI unit of radioactivity is the Becquerel (Bq). The Becquerel is defined as one disintegration per second. The relationship between curie and Becquerel is 1 curie = 3.7×10^{10} Becquerel.

Isotopes – atoms with same atomic number but different mass numbers
Radioactivity is the spontaneous decay of unstable elements through the emission of radiations.
Radioactivity is unaffected by external parameters such as temperature and humidity.
Unit of activity is Curie. SI unit is Becquerel.
Radioactive decay follows exponential law.

2.1.4 Radioactive Decay

On emission of an α or β particle, the element gets converted into a new element. This is the process of radioactive decay. The activity of any radioactive material depends on the concentration of radioactive atoms in it and not on the mass of the material present. For example, two one-curie sources of Cs-137 might have very different masses depending upon the relative proportion of non radioactive atoms present in each source. Each radionuclide decays at its own unique rate, which cannot be altered by any chemical or physical process. This process of radioactive decay is according to an exponential law, known as the radioactive decay law. Mathematically it may be stated as

$$N = N_o e^{-\lambda t} \qquad \text{... (2.1.2)}$$

where N_o is the number of radioactive atoms present at some reference time t=0, N is the number of radioactive atoms left after time t has passed and λ is called the disintegration or decay constant. λ is a characteristic of the radioactive material. Materials with higher values of λ decay rapidly and vice versa.

In actual practice, the decay of a radioactive isotope is usually given in terms of its half life denoted by $t_{1/2}$. The half-life is defined as the time required for the activity of any particular radionuclide to decrease to one-half of its initial value. The half life is a characteristic of a particular radioisotope and is different for different radioisotopes. It varies from fraction of a second to millions of years. In radiography we use isotopes with half life varying from a few days to years. Sources with shorter half life die out quickly. Substituting $N = N_0/2$ and $t = t_{1/2}$ the decay equation may be modified as

$$\frac{N_o}{2} = N_o e^{-\lambda t_{1/2}} \qquad \text{... (2.1.3)}$$

$$t_{1/2} = 0.693/\lambda \qquad \text{... (2.1.4)}$$

The decay equation shows that theoretically it requires an infinite time for the complete decay of a radioactive sample. When plotted graphically the decay equation results in what is known as a decay curve represented as a histogram in Fig. 2.1.2

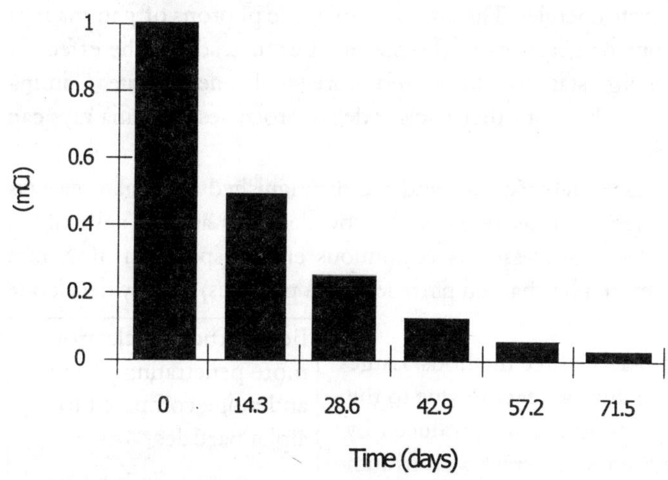

Fig. 2.1.2 Decay of a radioactive isotope

> Half Life – Time required for any radionuclide to decrease to one half of its initial value.
>
> Radioactive isotopes emit
>
> *Alpha particles*
>
> *Beta particles*
>
> *Gamma radiation*
>
> *Neutrons*
>
> Alpha particles –
>
> *positively charged particle with a charge of 2 and mass of four.*
>
> *Very small range.*

2.1.5 Properties of Radiation

A radioactive substance on decay usually emits one or more of the four types of radiations namely - alpha, beta, gamma or neutrons.

Alpha Particles – Certain radionuclides of high atomic mass (Ra226, U238, Pu239) decay by the emission of alpha particles. These alpha particles are positively charged and consists of 2 protons and 2 neutrons. It is thus identical to the nucleus of the helium atom. Alpha particle is not very penetrating and almost all the alphas can be stopped by a thin sheet of paper. The nucleus that emits alpha particle loses two protons and two neutrons. Therefore, after disintegration the atomic number of the atom is reduced by 2 and its mass number by 4.

e.g. $$^{226}Ra_{88} \longrightarrow {}^{222}Rn_{86} + {}^{4}He_{2}$$

Alpha particles are emitted with discrete energies characteristic of the particular transformation from which they originate. All alpha particles from a particular radionuclide transformation will have identical energies.

Beta Particles – A nucleus with a slightly unstable ratio of neutrons to protons may decay through the emission of a high speed electron called a beta particle. When a beta is emitted, the atomic number increases by one and the mass number remains unchanged.

e.g. $$^{3}H_{1} \longrightarrow {}^{3}He_{2} + {}^{0}e_{-1}$$

The beta particles emitted by a specific radionuclide range in energy from near 0 up to a maximum value characteristic of the particular transformation. The beta rays are more penetrating than alpha rays; but most beta particles can be stopped by few millimeters of Aluminium.

Gamma-rays – A nucleus which is in an excited state may emit one or more photons (particles of electromagnetic radiation) of discrete energies. The emission of these photons of gamma rays does not alter the number of protons or neutrons in the nucleus but instead has the effect of moving the nucleus from a higher energy state to a lower and more stable energy state. Gamma ray emission frequently follows beta or alpha or other nuclear decay processes. Gamma rays can penetrate through several centimeters of lead.

X-rays are also part of the electromagnetic spectrum and are distinguished from gamma rays only by their origin. X-rays are emitted with discrete energies by electrons as they shift orbits following certain types of nuclear decay processes. A continuous energy spectrum of X-rays called bremsstrahlung may also be emitted by charged particles (beta particles) as they decelerate near atomic nuclei.

Neutrons are typically produced by one of three methods. Large amounts of neutrons are produced in nuclear reactors due to the nuclear fission process. High energy neutrons are produced by accelerating deuterons and causing them to interact with tritium nuclei. The third method of producing neutrons is by bombarding beryllium with alpha particles. Neutron sources can be made using the alpha-neutron reaction on beryllium by making a mixture of powered alpha emitter and beryllium and sealing it in a metal container. Early neutron sources used radium as the alpha emitter. Modern neutron sources typically use plutonium or americium as the alpha source. The radium-beryllium (Ra-Be) sources were also sources of large amounts of gamma radiation while the plutonium-beryllium (Pu-Be) sources

> Beta particles – electrons, more penetrating power and range compared to alpha particles
>
> Gamma rays emitted by excited nucleus
>
> X-rays are emitted due to transition of electrons within orbit (characteristic X-rays) and deceleration of charged particles near an atomic nuclei. (Bremstrahlung)

and the americium-beryllium (Am-Be) sources only produce small amounts of very low energy gamma radiation. Thus, as neutron sources, Pu-Be and Am-Be sources tend to be less hazardous to handle. The older Ra-Be sources also had a tendency to develop leaks over time and give off radon gas, one of the products of radium decay. Sb-Be is also one of the sources, but it has a high gamma background.

2.1.6 Energy of Radiation

Energy of radiation is usually specified in terms of 'Electron Volt'. An electron volt is defined as the amount of kinetic energy acquired when an electron moves across a potential difference of one volt. Thus, if an electron moves across a potential difference of 1 V it acquires kinetic energy is equivalent to 1 eV. The energy gained by an electron moving through a potential difference of 1 V can also be expressed in terms of joules as

$$1 \text{ electron volt (1 eV)} = 1.6 \times 10^{-19} \text{ joules}$$

In the case of particulate radiation, the energy is the kinetic energy of motion $\frac{1}{2}mV^2$, while in the case of X- and gamma radiations it is the radiant energy or photon energy ($h\nu$).

Alpha particles and gamma radiation are emitted with energies. On the other hand, beta particles emitted by radioisotopes can have energies from 0 to a maximum energy E_{max} characteristic of the radioisotope. The energy of the beta particles emitted by a radioisotope is usually specified in terms of the maximum energy. The average energy is normally 1/3 of the maximum.

2.1.7 Nature of X- and Gamma Rays

X-rays and gamma-rays are forms of electromagnetic radiation having wavelengths roughly in the range 10^{-10} to 10^{-7}cm. X-rays are produced by allowing a stream of high energy electrons to strike a metal target, and they originate in the extra-nuclear structure of the target atoms. Gamma-rays are emitted from the nucleus of radioactive elements. The essential difference between light, ultra-violet, infra-red, radio-waves, and X-rays is one of wavelength, or frequency.

Being electromagnetic radiations, X- and gamma rays exhibit similar properties. In brief these are:

i. They are invisible
ii. They cannot be felt by human senses
iii. They cause certain materials to fluoresce. The fluorescent materials are zinc sulfide, calciumtungstate, diamond, barium platinocyanide, naphthalene, anthracene, stilbene, thallium activated sodium iodide, etc.
iv. They travel at the speed of light i.e. 3×10^{10}cm/sec
v. They are harmful to living cells.
vi. They can cause ionization.

> Energy of radiation given in terms of electron volt (eV).
>
> 1 eV is the energy gained by an electron in moving across a potential difference of 1 V.
>
> X-rays and gamma rays are electromagnetic radiation of very short wavelength.

vii. They travel in straight lines.

viii. They obey the inverse square law

ix. They can penetrate materials. Penetration depends upon the energy of radiation and the range depends on the density and thickness of the material.

x. They affect photographic emulsions.

Properties (iii), (vii), (viii), (ix), (x) are used in industrial radiography.

2.1.8 Inverse Square Law

The intensity of radiation reaching a point is governed by its distance from the source. The intensity varies inversely with the square of this distance. This is known as the inverse square law of radiation. This principle is illustrated in Figure 2.1.3

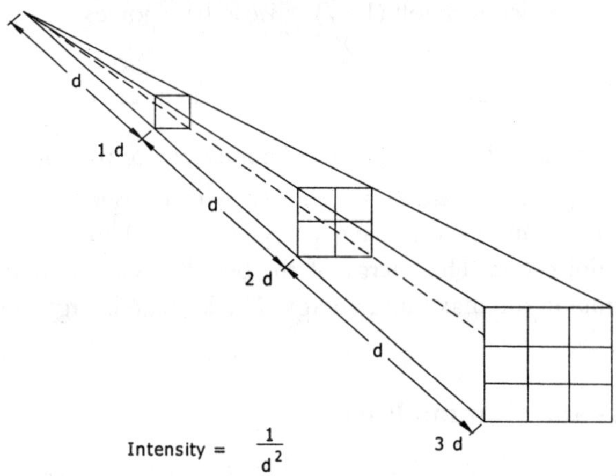

Fig. 2.1.3 Inverse square law

Stated algebraically :

$$\frac{I_1}{I_2} = \frac{(d_2)^2}{(d_1)^2} \qquad \text{... (2.1.5)}$$

where I_1 is the intensity at a distance d_1 and I_2 is the intensity at a distance d_2.

The inverse square law has important implications in practical radiography. The film should receive a certain amount of radiation dose or exposure to produce a radiograph of a particular density. If due to some reason the distance between the source and the film has to be altered then the exposure should be modified in accordance with the inverse square law. The inverse square law plays a vital role in radiation protection as well. The further away a person is from the source the lesser would be the dose received by him.

Example : If the radiation intensity at a point 4 m from the source is 100 mR/hr, what will the intensity be at 8 m.?

Solution : Eqn : 2.1.5 can be rewritten as $I_2 = \frac{(d_1)^2}{(d_2)^2} \times I_1$

Substituting I_1 = 100 mR/hr, d_1 = 4 m and d_2 = 8m, we have
$I_2 = 100 \times 4 \times 4/(8 \times 8) = 25$ mR/hr

2.1.9 Radiation Intensity and Specific Emission

The intensity of radiation may be defined as the number of rays per second falling perpendicularly over a unit area. The radiation output of a given radioactive source is measured in terms of roentgens per hour at a distance of 1 meter from the source per curie. This is known as the RHM value of the source. Each radioactive source has a specific RHM value per curie. Thus RHM for Cobalt 60 is 1.3 R/hr and for Iridium 192 it is 0.5 R/hr. The number of roentgens per hour at 1 cm from a 1 mC source of a gamma-ray emitter is called the K – factor.

The roentgen itself may be defined as that amount of X or gamma radiation which when it passes through 1 cm^3 of dry air at N.T.P. (Weighing 0.00129 gm) produces ions equivalent to one e.s.u. of charge of either sign. Roentgen is also equivalent to energy absorption of 87.7 ergs/gm of the exposed material.

The specific activity of a radioisotope source is defined as the activity per gram. It is usually measured in curies per gram. A higher specific activity means that a source of a given strength can be produced in a smaller physical size, which is of great importance from the point of view of image sharpness. Also a small dimensioned source has less self absorption of radiation and so a greater effective output. The specific activity depends on the reactor flux, time for which the material is irradiated, as well as on the characteristics of the irradiated materials such as the atomic weight and the activation cross-section. Some elements can be activated to a very high specific activity while others are incapable of reaching high activities with available neutron fluxes.

> Radiation output of given source measured in terms of roentgen per hr at 1 m (RHM).
>
> Each isotope has a characteristic RHM value
>
> Specific activity is the activity per unit gram
>
> Higher specific activity
> - *Smaller is the source*
> - *Better is the image sharpness*
> - *Lesser is the self absorption*
>
> Specific activity depends on
> - *Neutron flux and characteristics of material irradiated.*

Chapter 2.2
Interaction of Radiation with Matter

As the nuclear radiations move through matter, they lose their energy through various interactions with the atoms of the material. The rate at which the energy loss occurs depends upon the type and energy of radiation and composition and density of the material through which it passes.

2.2.1 Ionisation and Excitation

The three main types of radiation namely alpha, beta, X- and gamma radiation lose their energy primarily through excitation and ionization of orbital electrons. The term "excitation" is used to describe an interaction between the incident radiation and orbital electron of an atom of the material where the electron acquires energy from a passing charged particle but is not removed completely from the atom. Excited electrons may subsequently emit energy in the form of X-rays during the process of returning to a lower energy state. "Ionization" refers to the complete removal of an electron from an atom following the transfer of energy from a passing charged particle or photon. An atom is electrically neutral. When an electron is removed from an atom, it is left with positively charged ion. The expelled electron may attach itself to another atom, thus creating a negative ion. To remove an electron from an atom, a certain minimum energy is needed to be imparted to the atom, which is called the ionization energy. Production of an ion pair i.e. a positive ion and a negative ion in air needs an energy of about 34 electron volts. In describing the intensity of ionization, the term "specific ionization" is often used. This is defined as the number of ion pairs formed per unit path length for a given type of radiation.

> Ionisation – *Process of creation of an ion*
>
> Excitation – *Raising an electron to a higher energy state*
>
> Alpha particles
> * *Interact principally by ionization and excitation*
> * *Range is small*
> * *High specific ionisation*
>
> Beta particles lose energy through
> * *Ionization, excitation and bremstrahlung*
> * *Higher range but lower specific ionisation compared to alpha particles*

2.2.2 Interaction of Alpha Particles

When alpha (α) particles and other heavy ions move through matter, they interact mainly by ionisation and excitation of the atoms of the absorber. These are called collision processes. These processes lead to a gradual energy loss until the particle comes to rest after traversing a certain thickness of material called the range. Because of their double charge and relatively slow velocity, alpha particles have a high specific ionization ranging from about 20,000 to 80,000 ion pairs per cm and relatively short range in matter. The range of alpha particles is about 2 inches in air. They can be stopped completely by a thin sheet of paper.

2.2.3 Interaction of Beta Particles

The interaction of beta particles is very similar to that of alpha particles. However, for a given energy, the velocity of an electron is much larger than that of an alpha particle. Hence the specific ionisation i.e. the number of ion pairs produced per millimetre of beam path is less compared to alpha particles. For beta particles of high energy, an additional mechanism of losing energy must be taken into account–the emission of radiation (bremsstrahlung) when an electron is accelerated/decelerated in the electric field of the nucleus. The maximum distance a beta particle can penetrate in matter is defined as its range (R). The range of beta particles depends on its energy. In general, beta particles can be stopped by a thin sheet of aluminium.

> Neutrons lose energy through moderation and interaction with the nucleus
>
> X-rays and gamma rays interact through
> * *Photoelectric effect*
> * *Compton scattering*
> * *Pair Production*

2.2.4 Interaction of Neutrons

The interaction of neutrons is different from that of charged particles and gamma rays. The chief processes are scattering and absorption by the nucleus by virtue of the nuclear forces alone. Neutrons lose energy in matter by collisions, which transfer kinetic energy. This process is called moderation and is most effective if the matter the neutrons collide with has about the same mass as the neutron. Once slowed down to the same average energy as the matter being interacted with (thermal energies), the neutrons have a much greater chance of interacting with a nucleus. Such interactions can result in material becoming radioactive or can cause radiation to be given off.

2.2.5 Interaction of X- and Gamma Rays

X-rays and gamma rays are electromagnetic radiations, which attenuate in matter exponentially and do not have definite range as is found in charged particles. When a beam of X- or gamma ray photons are incident on a thin absorber, each photon is removed from the beam in a single event. This event may be an actual absorption process in which case the photon disappears or the photon may be scattered out of the beam. Fig. 2.2.1 is a typical sketch indicating the range of alpha beta and gamma radiations. Three principal processes by which X -or gamma ray photons interact with matter are

(a) Photoelectric effect,
(b) Compton scattering and
(c) Pair production.

These are described below

2.2.5.1 Photoelectric Absorption

The photoelectric process (Fig. 2.2.2) occurs when the incident photon (primarily of low energy) interacts with an inner orbital electron. When this reaction occurs, all the energy of the photon is transferred to the electron and the photon is completely absorbed. Since the electron is bound to the atom, part of the energy of the photon is used for overcoming this binding energy. The balance appears as the kinetic energy of the electron, which is ejected, from the atom. This

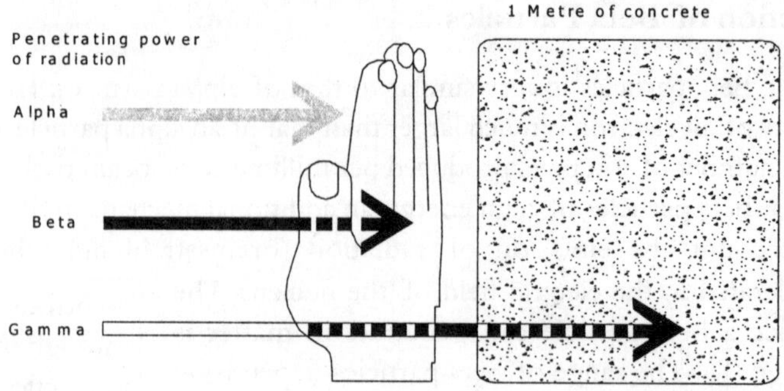

Fig. 2.2.1 Schematic sketch indicating the range of α, β and γ radiations (Reproduced with permission from http://www.ndt-ed.org/EducationResources/CommunityCollege/Radiography)

electron is also referred to as photoelectron. Since a vacancy now exists in the electron shell, other orbital electrons will fall into that vacancy with a subsequent emission of characteristic X-rays. Photoelectric effect occurs at predominantly low energies (upto about 150 keV). It also increases with increasing atomic number.

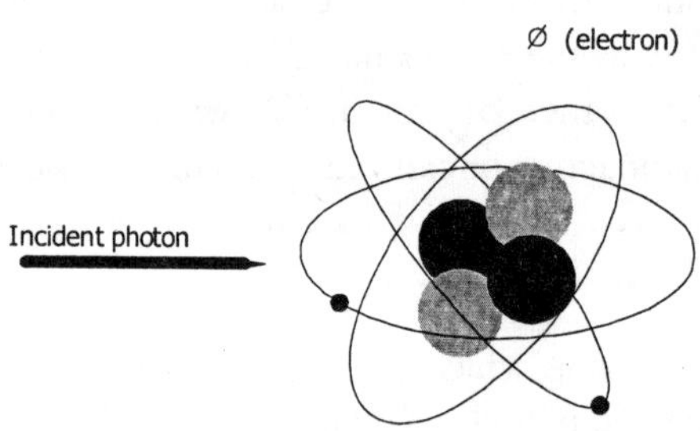

Fig. 2.2.2 Photoelectric effect

Photoelectric effect
Incident photon absorbed and photoelectron liberated
Predominant at low energies <150 keV
Increases with increasing atomic number
Compton scattering - incident photon transfers partial energy to electron
Electron ejected from atom and photon moves in different direction with reduced energy
Predominant in medium energy range
Contributes to scattering thus degrading image quality in industrial radiography

2.2.5.2 Compton Scattering

Compton scattering also called as incoherent scattering occurs when an incident photon (primarily of medium energy) interacts with a free or loosely bound electron of the outer shell. During this interaction, a portion of the energy of the incident photon is transferred to the electron. The amount of energy transferred depends on the angle at which the photon collides with the electron. If the collision is head-on, maximum energy is transferred. If the collision is of glancing nature, minimum energy is transferred to the electron. As a result of the collision, the electron (also called as Compton electron) is ejected from the atom while the photon emerges from the atom with reduced energy and changed direction (Fig. 2.2.3). In industrial radiography, Compton scattering contributes to reduction in the radiographic image quality. Compton scattering is predominant in the medium energy range from about 150 keV to about 3 MeV.

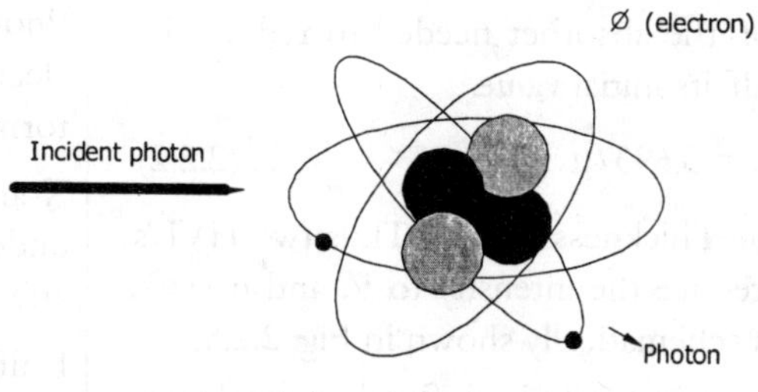

Fig. 2.2.3 Compton Scattering

2.2.5.3 Pair Production

The third mechanism by which X or gamma photons interact with matter is pair production. This phenomenon occurs only when the energy of the photon exceeds 1.02 MeV. The high energy photon is completely absorbed. In its place a electron positron pair is produced (Fig. 2.2.4). The positron annihilates immediately with an electron resulting in the emission of two gamma rays of 0.51 MeV. This is referred to as annihilation radiation.

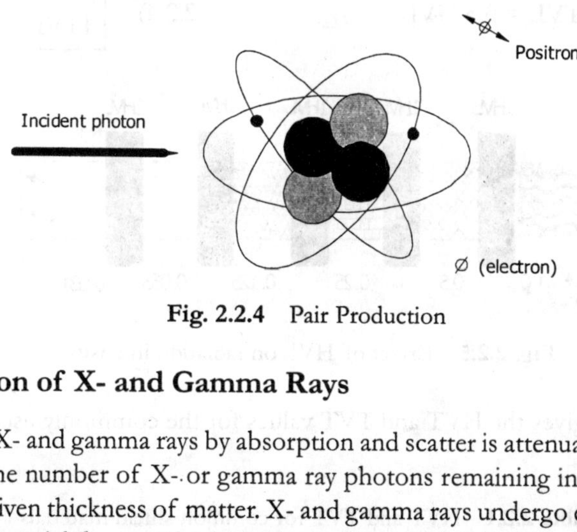

Fig. 2.2.4 Pair Production

2.2.6 Absorption of X- and Gamma Rays

The interaction of X- and gamma rays by absorption and scatter is attenuation. Attenuation is the total reduction in the number of X- or gamma ray photons remaining in the beam following its passage through a given thickness of matter. X- and gamma rays undergo exponential attenuation i.e. there is an exponential decrease in the intensity of radiations as a homogeneous beam of X- or gamma-rays pass through matter. If I_o is the intensity of incident X- or gamma rays and I is the intensity of transmitted X- or gamma rays, mathematically, the exponential decrease can be represented as,

$$I = I_0 e^{-\mu x} \qquad \text{... (2.2.1)}$$

where μ is the linear attenuation coefficient of the material (cm^{-1}) and x is the thickness of the material (cm).

2.2.7 Half Value layer and Tenth Value Layer

The attenuation of X- or gamma ray beam may also be expressed in terms of a quantity called the 'Half-Value Thickness'. This is defined as the thickness of the absorber needed to reduce the intensity of radiation to half its initial value.

$$x_{1/2} = 0.693/\mu \qquad ... (2.2.2)$$

where $x_{1/2}$ is the Half-Value Thickness (HVT). Thus two HVL's of shielding material will reduce the intensity to ¼ and n-HVL will reduce to $1/2^n$. This is schematically shown in Fig. 2.2.5.

Tenth value layer ($x_{1/10}$) is similarly defined as thickness necessary to reduce gamma ray intensity to 1/10 of its initial value.

$$x_{1/10} = 2.3/\mu \qquad ... (2.2.3)$$

Thus, 2 TVLs will reduce the intensity to $1/10^2 = 1/100$ of its initial value and n TVT will reduce to $1/10^n$ of the initial intensity and so on.

HVL and TVL are related by the equation

$$1 \text{ TVL} = 3.3 \text{ HVL} \qquad ... (2.2.4)$$

Pair production – occurs only when incident photon energy exceeds 1.02 MeV.

Photon disappears and electron positron pair formed.

X-and gamma rays undergo exponential attenuation in matter

Half value layer – thickness of the material required to reduce the intensity of radiation to half its original value.

Tenth value layer – thickness of the material required to reduce the intensity of radiation to one tenth its original value.

1TVL = 3.3 HVL

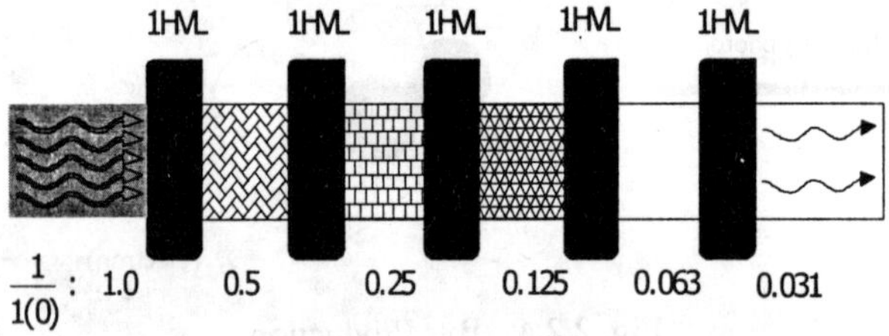

Fig. 2.2.5 Effect of HVL on radiation intensity

Table 2.2.1 below gives the HVT and TVT values for the commonly used radionuclides and shielding materials.

Table 2.2.1 HVT and TVT for common shield materials

Radionuclide	Material thickness (cm)							
	Concrete		Steel		Lead		Uranium (depleted)	
	HVT	TVT	HVT	TVT	HVT	TVT	HVT	TVT
Iridium–192	4.1	13.5	1.3	4.3	0.6	1.9	0.3	1.1
Caesium–137	4.9	16.3	1.5	5.0	0.7	2.2	0.4	1.2
Cobalt–60	6.1	20.3	2.0	6.7	1.2	4.0	0.7	2.2

Chapter 2.3
Radiation Detection

> Radiation can be detected using
>
> *Film*
>
> *Gas Filled Detectors*
>
> *Semiconductor detectors*
>
> Gas filled detectors based on principle of ionization of gas.
>
> Gas used air or argon

Since we cannot see, smell or taste radiation, we are dependent on instruments to indicate the presence of ionizing radiation.

The most widely used medium for detection is the film. The basic advantage of industrial radiographic films is its ability to provide a visual two dimensional image. The resolution of the film is quite high making it possible to record the minute details in the objects such as the presence of cracks and other defects very effectively. Film is treated in more detail in Section 4. This Chapter briefly reviews the detection of ionising radiation using other detectors such as ionisation chambers and scintillation detectors. Apart from these we also have semiconductor detectors such as Si(Li) and Ge detectors which find widespread applications for X- and gamma ray spectrometric applications.

2.3.1 Gas Filled Detectors

The most common type of instrument is a gas filled radiation detector. Typically the detector consists of a closed chamber filled with gas usually air or argon and provided with two electrodes as shown in fig 2.3.1.This instrument works on the principle that as radiation passes through an air or a specific gas between the electrons, ionization of the molecules air occur. When a high voltage is applied inside the chamber, the positive ions will be attracted to the negative side of the detector (the cathode) and the free electrons will travel to the positive side (the anode). These charges are collected by the anode and cathode, which then form a very small current in the wires going to the detector. By placing a very sensitive current measuring device between the wires from the cathode and anode, the small current can be measured and displayed as a signal. The more the radiation entering the chamber, the more current would be displayed by the instrument.

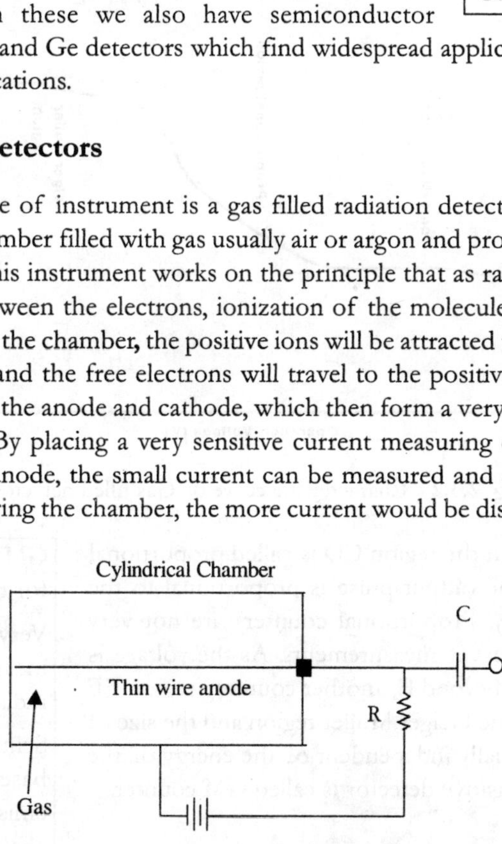

Fig. 2.3.1 Gas filled ionization type detector

Fig. 2.3.2 is useful in understanding the operation of various instruments. Depending upon the potential difference applied to the ion chamber many types of gas-filled detectors exist, but the three most common ones are the ionization chamber, proportional counter and the GM detector. When radiation enters the chamber, it ionises gas producing electron ion pairs Under the influence of the applied voltage, the electrons are attracted towards the anode and the positive ions towards the cathode. A current flows in the circuit. As the applied voltage is increased, more ion pairs are collected as shown in fig. 2.3.2. But after a certain voltage, the current does not increase with further increase in voltage because all ions produced in the chamber are collected (point B). This voltage is known as the saturation potential and the corresponding current is called the saturation current. Ionisation chamber detectors operate below this saturation voltage. The number of electrons collected is equal to the number of primary electrons produced. There is no amplification in the ionization chambers. The saturation current is directly proportional to the strength of the radiation. A higher radiation level results in increase in the current. The ionisation chamber detector is not very very sensitive, as there is no amplification. The ionization chamber is used for measuring large amounts of radiation. The teletector and the pocket dosimeter are typical examples of ion chamber instrument.

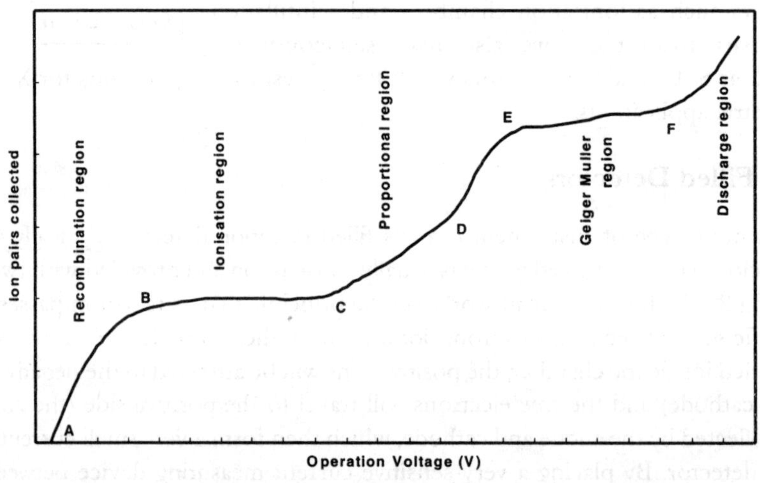

Fig. 2.3.2 Characteristic curve of Gas filled detector

A detector operated in the region CD is called proportional counter as the size of the output pulse is proportional to the original ionization energy. Proportional counters are not very much used in radiation survey measurements. As the voltage is increased still further i.e., beyond E, another counting region EF is reached. This is called the Geiger Muller region and the size of the output pulse is essentially independent of the energy of the initial ionization. This sensitive detector is called GM counter.

> GM counters widely used for radiation survey.
>
> Very sensitive and can measure small amounts of radiation
>
> Scintillation detectors based on the principle of emission of light on absorption of radiation.

The principle of operation of GM counter is basically the same as the ionization chamber, except that the voltage across the chamber is much higher. Because of this high voltage, electrons produced due to ionization are accelerated towards the anode rapidly. They collide with gas atoms on the way and produce secondary electrons, which in turn produce still more electrons. Thus each original electron created by the initial ionizing event produces an avalanche of electrons. With the increased voltage the ionizing event is magnified by the avalanche to such an extent that the size of the pulse is no longer a function of number of ions formed in the initial event. GM counters are very sensitive and can measure very small amounts of radiation. They are rugged and compact and thus ideally suited for field applications.

2.3.2 Scintillation Detectors

The second most common type of radiation detecting instrument is the scintillation detector. The basic principle behind this instrument is the use of a special material, which glows or "scintillates" when radiation interacts with it. The most common type of material is a type of salt called sodium-iodide. The light produced from the scintillation process is reflected through a clear window where it interacts with device called a photomultiplier tube.

The first part of the photomultiplier tube is made of another special material called a photocathode. The photocathode has the unique characteristic of producing electrons when light strikes its surface. These electrons are then pulled towards a series of plates called dynodes through the application of a positive high voltage. When electrons from the photocathode hit the first dynode, several electrons are produced for each initial electron hitting its surface. This "bunch" of electrons is then pulled towards the next dynode, where more electron "multiplication" occurs. The sequence continues until the last dynode is reached, where the electron pulse is now millions of times larger then it was at the beginning of the tube. At this point the electrons are collected by an anode at the end of the tube forming an electronic pulse. The pulse is then detected and displayed by a special instrument. This process is illustrated in fig 2.3.3.

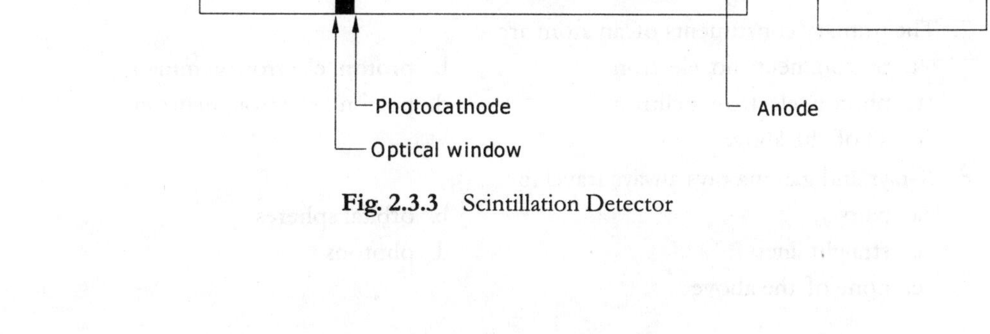

Fig. 2.3.3 Scintillation Detector

Scintillation detectors are very sensitive radiation instruments and are used for special environmental surveys and as laboratory instruments. Apart from scintillator detector we also have semiconductor detectors of germanium and Silicon which are widely used for gamma spectrometry.

Questions

Choose the most appropriate answers

1. One Curie is equivalent to :
   a. 100 millicurie
   b. 1000 microcurie
   c. 1000 millicurie
   d. 10,000 microcurie
   e. none of the above
2. The time required for one-half of the atoms in a radioactive material to disintegrate is referred to as :
   a. the inverse square law
   b. a curie
   c. half-life
   d. the exposure time
   e. decay constant
3. The term R/hr refers to
   a. Roentgen eqvt. man/hr
   b. Roentgens per hour
   c. Activity per hour
   d. Dose rate per hour
   e. None of the above
4. The widely used unit of measurement for measuring the rate at which a gamma ray source decays is the:
   a. Curie
   b. Half life
   c. MeV
   d. Roentgen
   e. Tenth life
5. Of all the radiations listed below, the one with shortest wavelength is
   a. Visible Light
   b. Ultra Violet
   c. 100 kV peak X-rays
   d. Infrared Radiation
   e. Electromagnetic radiation
6. Beta particles are
   a. fast moving charged particles
   b. protons
   c. electrons
   d. positrons
   e. neutrons
7. The primary constituents of an atom are
   a. proton, neutrino, electron
   b. proton, electron, gamma ray
   c. photon, electron, neutron
   d. proton , electron, neutron
   e. all of the above
8. X-rays and gamma rays always travel in
   a. pairs
   b. orbital spheres
   c. straight lines
   d. photons
   e. none of the above

9. The term used to describe the loss of excess energy by the nucleus of radioactive atom is called :
   a. decay (disintegration)
   b. ionization
   c. scintillation
   d. activation
   e. excitation
10. X-rays, gamma rays, and alpha particles all have one thing in common; they are all :
    a. particulate radiations
    b. electromagnetic radiations
    c. microwave radiations
    d. ionizing radiations
    e. all of the above
11. The term used to express the number of curies of radioactivity per gram is :
    a. decay activity
    b. specific activity
    c. M factor
    d. Atomic weight
    e. None of the above
12. Particles that carry either a positive or negative electrical charge are referred to as :
    a. photoelectrons
    b. photons
    c. ions
    d. protons and electrons
    e. ionising particles
13. Gamma and X-radiations interact with matter and may be absorbed by :
    a. Photoelectric absorption
    b. Compton scattering
    c. Pair production
    d. Any of the above
    e. None of the above
14. Approximately what percent of the original radioactivity is left after 4 half-lives ?
    a. 50
    b. 6.25
    c. 12.5
    d. 25
    e. None of the above
15. The intensity of radiation emitted by a radioisotope :
    a. increases with time
    b. decreases with time
    c. is a constant irrespective of time
    d. depends on external parameters
    e. any one of the above

SECTION 3

SOURCES FOR INDUSTRIAL RADIOGRAPHY

After reading this section, you should be able to answer

- ★ *How X-rays and gamma rays are produced*
- ★ *What are the advantages and limitations of X- and gamma ray sources*
- ★ *Explain the construction and working of an X-ray machine*
- ★ *List the sources for gamma radiography with their characteristic features*
- ★ *Differences in quality of X-and gamma ray emissions from their source*

Chapter 3.1
X-Ray Equipment

3.1.1 Introduction

X-rays are produced when fast moving electrons collide with matter or are suddenly brought to rest. The essential requirements for the production of X-rays include (a) source of electrons, (b) means for directing and accelerating the electrons and (c) a target aligned with the electron beam for the electrons to strike and produce X-rays. Heating a filament, ionisation of gas, or field emission from a point source may produce electrons. In conventional X-ray tubes used for medical and industrial applications, heating the filament produces electrons. A high voltage applied to the tube accelerates the electrons on to the target. The sudden stopping of these electrons in the target results in the generation of X-rays. Conventional X-ray equipment thus consists of three main components (a) X-ray tube, (b) high voltage unit and (c) control unit. This is illustrated as a simple schematic in Fig. 3.1.1

X-ray machine consists of

* *X-ray tube*
* *HV unit*
* *Control unit*

X-ray tube houses

* *cathode "-" vely charged electrode containing filament*
* *Anode "+"vely charged electrode containing target*

X-ray tubes can be made of

* *Glass or*
* *Metal Ceramic*

Vacuum in X-ray tube

* *prevents oxidation*
* *Permits free flow of electrons*
* *provides electrical insulation between anode and cathode*

3.1.2 X-Ray Tube

An X-ray tube consists of a highly evacuated glass bulb into which are sealed two electrodes the cathode and anode (Fig. 3.1.2). The cathode is the negatively charged electrode containing the filament. The anode is the positively charged electrode in which the target for producing X-rays is embedded. Electrons produced by heating the filament are accelerated by applying a high voltage across the anode and cathode and then stopped suddenly by the target resulting in the generation of X-rays.

3.1.2.1 Envelopes

The tube envelope may be made of glass or it can be of the metal ceramic type. The glass is normally of the 'Pyrex' variety or borosilicate 'hard' glass type. Glass tubes though still widely used are susceptible to thermal and mechanical shocks. Most of the modern industrial X-ray units use metal ceramic tubes. Metal ceramic tubes are compact, provide efficient insulation and greater shock resistance and hence are more rugged compared to gas envelope tubes. The envelope is evacuated and maintains pressure less the 10^{-6} mm Hg. A high vacuum environment is necessary to

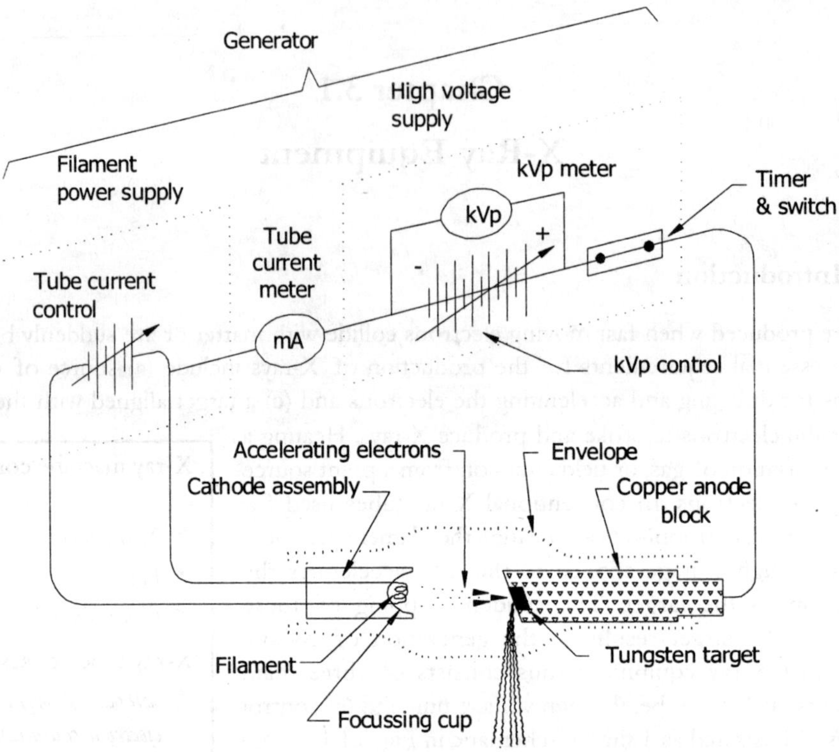

Fig. 3.1.1 Simplified schematic sketch of an X-ray equipment

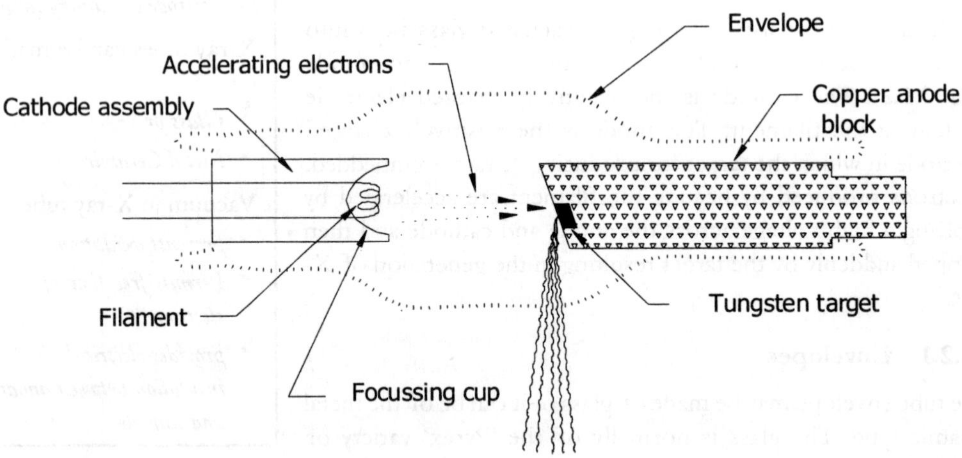

Fig. 3.1.2 Simplified sketch showing the major interior elements in an X-ray tube

1. prevent oxidation of the filament
2. permit free flow of electrons within the tube.
3. provide electrical insulation between the electrodes

3.1.2.2 Cathode

The cathode assembly consists of one or more filament (only one of which is used at a time) and the focusing cup. When current of the order of 3-10 amperes is passed through the filament, it becomes hot and emits electrons (thermionic emission). Tungsten is the commonly used filament material.

Because of the large number of electrons and the forces of mutual repulsion between them , the electrons emitted from the filament tend to spread itself out. This is prevented by the cathode focusing cup. The focusing cup is a shaped metal electrode usually made of nickel. When the X-ray tube is conducting, the focusing cup is maintained at the same negative potential as the filament. The focusing cup is designed so that its electrical forces cause the electron stream to converge onto the target anode in the required size and shape. The tube current passing between the cathode and anode ranges from microamperes in the case of microfocal units to more than 15 milliamperes in industrial X-ray units.

Cathode consists of
* *Qualified*
* *Filament*
* *Focusing cup*

Electrons are produced by thermionic emission – heating of filament

Anode consists of
* *Target*

Efficiency of X-ray production depends on
Voltage applied
Atomic number

About 99% of the energy of electrons is converted to heat.

Target should have
* *High Z*
* *High melting point*
* *Low vapour Pressure*
* *High thermal conductivity*

Tungsten most widely used target

3.1.2.3 Anode

The anode consists of a metal electrode of high thermal conductivity containing the target. Electrons from the filament are accelerated and impinge on the target to produce X-rays. In conventional X-ray units, about 99 % of the kinetic energy of the impinging electrons appears as heat and only about 1 % is converted into X-rays. The efficiency of X-ray production (E) is a function of accelerating potential (V) and atomic number of the target (Z) and is given by

$$E = k.V.Z \quad \text{... (3.1.1)}$$

where k is a constant. Thus, the target should possess the following properties

(a) It should have a high melting point
(b) It should have a high atomic number (Z)
(c) It should have a high thermal conductivity to dissipate the heat quickly and a low vapour pressure (to prevent evaporation of the target material).

In industrial X-ray units the widely used target is a tungsten button set in a block of vacuum cast copper. Tungsten has three physical attributes that make it the target material of choice

(a) It has a high melting point (3380°C). It can therefore withstand the high local temperatures produced by intense electron bombardment. Thus, higher intensity of X-radiation is possible.

(b) It has a high atomic number (Z=74)

(c) It has a low vapour pressure

Apart from tungsten, platinum and gold can also be used as target materials. Since most of the energy of electrons is lost as heat, the joint between copper and target should be good to ensure good and quick conduction. A suitable coolant such as oil can also be used for cooling the anode.

The focal spot is the region on the anode where the electrons impinge and from which X-rays are emitted. Many machines have dual focus – a smaller focus and large focus. Smaller focal spots are preferred, since they produce sharp radiographic images. But small focal spot means reduced X-ray intensity. Further, the intense heat generated, can vaporize the target, thus reducing tube life. To deal with this situation, the target is oriented between 0 to 30° degrees with respect to the vertical so that the projection of the electron beam spot on the target appears smaller when viewed from the radiographic direction. That is the actual focal spot would be larger while the apparent focal spot would be smaller. This is the principle of line focusing and is shown in Fig. 3.1.3. A target inclination angle of 20° is common. Measurement of focal spot size is normally done using the pinhole method.

> Focal spot – *point on anode from where X-rays originate*
>
> Target *inclined upto 20 deg. to decrease effective focal spot (principle of line focusing)*
>
> Focal spot measured by *pinhole method*
>
> Intensity of X-ray beam max at about 12 deg. From central ray *(heel effect)*

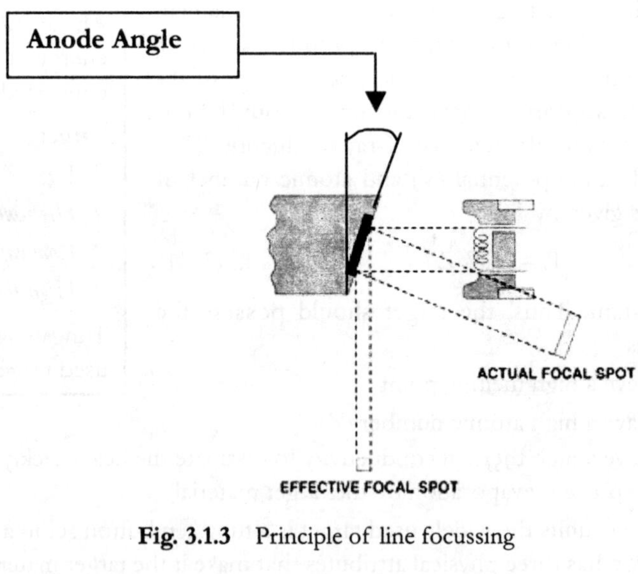

Fig. 3.1.3 Principle of line focussing

3.1.2.4 Heel Effect

The intensity of the X-ray beam that leaves the X-ray tube is not uniform throughout all portions of the beam. The intensity of the beam depends on the angle at which the X-rays are emitted from the focal spot. Assuming a target inclination of 20°, the maximum intensity occurs at about 12° from the central ray, and is about 5% higher. This variation is termed the "heel effect". This is shown in Fig. 3.1.4

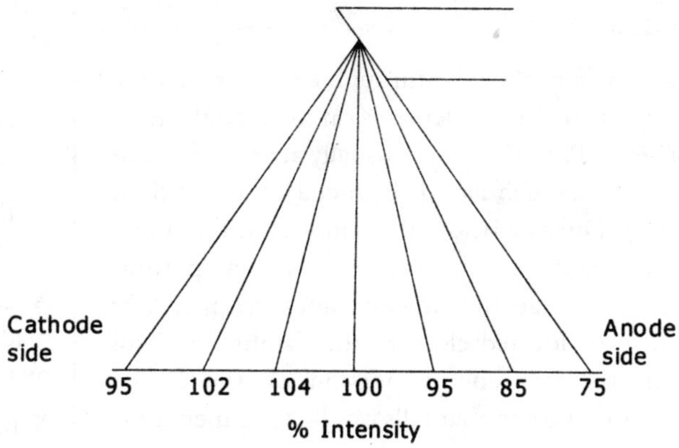

Fig. 3.1.4 Heel effect

3.1.2.5 Hood

Most of the modern X-ray tubes are hooded. This hood consists of a metal body over the anode with an aperture to permit the electron beam to reach the anode and another aperture for the X-ray beam (Fig 3.1.5). The hood is made of copper or tungsten and serves a number of purpose.

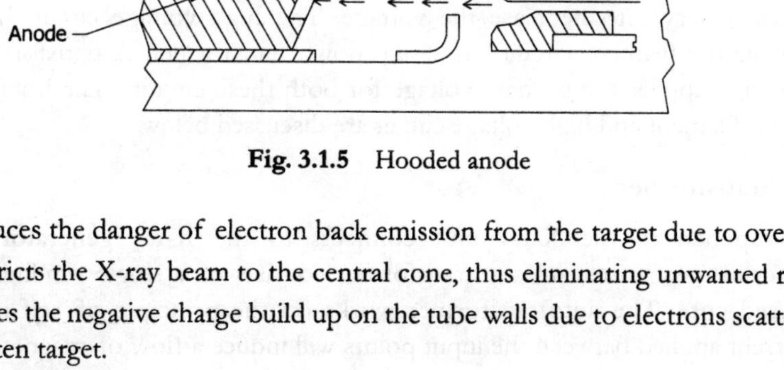

Fig. 3.1.5 Hooded anode

(a) it reduces the danger of electron back emission from the target due to over heating.
(b) it restricts the X-ray beam to the central cone, thus eliminating unwanted radiation.
(c) reduces the negative charge build up on the tube walls due to electrons scattered from the tungsten target.

3.1.3 Tubehead

The entire tube is enclosed in a metal casing lined with sufficient lead to restrict the leakage radiation. Tubeheads usually contain the X-ray tube, insulating oil or gas. In addition, the metal case is earthed to eliminate possible electrical hazards due to the high voltage used. The different type of tubeheads include unipolar tubes, bipolar tubes and rod anode type of machines.

3.1.3.1 Tube Window

The X-ray beam emerges from the tube through a port or window. The solid angle of the radiation cone depends on the manufacturer and varies from 20°-60°. The window is usually made of a less radiation absorbent material such as a light metal of low atomic number (e.g. beryllium). Directly below the window in the useful beam area is the diaphragm, a shielded opening that permits changing of the useful beam size. In many machines a copper filter may be incorporated in or located close to the diaphragm. This copper filter absorbs the low energy portion of the X-rays produced, making the useful beam that falls on the specimen more nearly uniform in energy.

Hooded anode prevents
* *electron back emission*
* *Restricts X-ray beam*
* *Reduces charge buildup on tube walls*

X-ray tube window made of low Z material (eg. Be)

X-ray Generator supplies power for
* *Heating filament to liberate electrons*
* *Accelerating electrons from cathode to anode*

3.1.4 X-Ray Generators

An X-ray generator is the device that supplies the necessary electrical power to the X-ray tube. The tube requires electrical energy for two purposes: to release electrons from the filament and to accelerate these electrons from the cathode to the anode. The X-ray generator has a circuit for each of these functions and we refer to them as the filament and high-voltage circuits (fig. 3.1.1).

The X-ray generator receives 230 V, 50 Hz (cycles per second) alternating current. Filament heating requires a potential difference of approximately 10 V, whereas electron acceleration requires a potential difference that needs to be varied between 10 kV and 450 kV. Transformers are used to change the potential difference of the incoming electric energy to the appropriate level. Transformers can be classified as step up or step down transformers. A step up transformer has more number of turns in the secondary coil and serves to increase the input voltage while a step down transformer serves to decrease the voltage. The high voltage circuit has a step up transformer while the filament circuit has a step down transformer. A transformer called the "autotransformer" supplies the primary voltage for both these circuits. The brief functions of autotransformer, filament and high voltage circuit are discussed below.

3.1.4.1 Autotransformer

The voltage supplied to the X-ray unit connects to the X-ray generator through an autotransformer in most cases. An autotransformer consists of a single winding wound on a laminated closed core. The autotransformer works on the principle of self-induction. An alternating current applied between the input points will induce a flow of magnetic flux around the core. This magnetic flux will link with all the turns forming the coil, inducing a voltage into each turn of the winding. The autotransformer has several functions. It provides the requisite voltage for the (a) filament circuit (b) primary of the high-voltage transformer and (c) other subsidiary circuits, which we will not consider.

3.1.4.2 Filament Circuit

The filament circuit regulates current flow through the filament of X-ray tube. The filament is a coiled tungsten wire that emits electrons when it is heated by this current flow. A current flow of 3 to 5 A with an applied voltage of about 10V are typical values. The power to heat the X-ray tube filament is provided by a small step-down transformer called the "filament transformer". The filament is connected directly to the secondary winding of this transformer. The primary winding of the filament transformer obtains its voltage by tapping off an appropriate number of turns from the autotransformer. Precise control of filament heating is critical, because a small variation in filament current results in a large variation in X-ray tube current.

Autotransformer
* *Based on the principle of self induction.*
* *Provides voltage for all circuits in X-ray unit.*

Filament Circuit
* *Step down transformer*
* *Provides current for heating filament*

HV Circuit consists of
* *Step up ransformer*
* *provides voltage for acceleration of electrons*

3.1.4.3 High Voltage Circuit

The high voltage circuit has two transformers, an autotransformer and a step-up transformer. The autotransformer is actually the kVp selector and is located in the control panel. The voltage across the primary coil of the step-up transformer can be varied by selecting the appropriate number of turns in the autotransformer. The step-up transformer, which is sometimes called the high-voltage transformer, has many more turns in the secondary coil than in the primary coil, and it increases the voltage by a factor of approximately 600. The potential difference across the secondary coil may be as high as 450 kV, so the step-up transformer is immersed in oil in the transformer assembly for maximum insulation. A voltmeter and an ammeter are incorporated into the high voltage circuit, to measure the applied voltage (kVp) and to measure the tube current (mA). The meters are located on the control panel, but their connections are in the high voltage circuit.

3.1.5 Control Panel

The control panel of an X-ray machine is designed to permit the operator to control the generation of X-rays so that exposures can be made in a simple and rapid manner. The panel also provides protective electrical circuits that prevent damage to the equipment. Depending on the complexity of the equipment and the electrical circuitry design, the control panel will comprise of a few or all of the following controls and indicators:

a. ***High Voltage Control and Meter:*** The high voltage control permits variation in the voltage (kV) applied across the tube. A voltmeter, usually calibrated in kilovolts, displays the voltage that has been set. Continuously variable voltages are possible in many of the present day machines.

b. ***Tube Current Control and Meter:*** The tube current control permits adjustment of tube current to exact values. An ammeter, usually calibrated in millamperes displays the tube current. While many machines have independent control for setting the tube current, in some machines the tube current is automatically set depending on the applied voltage.

c. ***Exposure Time:*** A synchronous timing device used to time exposures.

d. ***Power ON-OFF Switch:*** Controls the application of power to the equipment. In some equipment, this switch supplies power to the tube filament only. An independent ON-OFF switch controls the application of power to high voltage.

e. **Indicator Lamps:** Provides visual indication of status of power supply, HT, etc.

f. **Focal Spot Selector:** Used with tubes having dual focal spots, this switch permits selection of desired focal spot.

3.1.5.1 Equipment Protective Circuits

The equipment protective circuits serve to protect the equipment against excess heat generation or oil/water flow failure etc. Three of the important circuits include :

a. ***Over temperature Thermostat:*** Installed in the tubehead, the thermostat functions to remove power from the equipment when the anode gets heated up excessively.

b. ***Pressurestat:*** Installed in the tubehead of equipment using gas for insulation, the Pressurestat functions to remove power from the equipment when the gas pressure is below safe values.

c. ***Flow Switches:*** Installed in the oil and water circulators of equipment cooled by these means, the flow switches function by removing power from the unit when the cooling system fails.

3.1.5.2 Protective Electrical Circuits

The protective electrical circuits tend to protect the equipment from excess voltage or current that can cause damage. These include overload circuit breaker and over current/over voltage protection circuits.

3.1.6 Types of X-ray Machine Circuits

There are many variation of circuits that can be used to apply the high voltage needed for X-ray generation. These include the self rectified circuit, full wave rectified circuit, Villard circuit, Graetz circuit and Greinacker circuit. The waveforms are important since they control the amount of time that the peak voltage is applied across the tube. The nature of the circuit depends on the type of machine and manufacturer. Sometimes a filter circuit is also provided that "smooths out" the voltage waves. The Grienacker circuit is almost universally used in modern constant potential equipment.

3.1.7 Duty Cycle

The duty cycle is generally expressed as a percentage of exposure versus total time. For example a 100% duty cycle means that the tube can be operated continuously. 50% means that it is necessary to provide equal rest time after each exposure and so on. The life of an average tube operating under conventional loads may be from several hundred to over one thousand hours. One major factor concerning the tube life is the effectiveness of its cooling. If it is overloaded, its life is considerably shortened, while its life is lengthened if the tube is under loaded. Also the tube life can be increased by preheating it (warm up) and controlling its operation every time it starts working again after an idle period.

Duty Cycle
Percentage of exposure . vs. total time

Quality or energy of X-rays depend on
Voltage applied (kV)

Quantity or intensity of X-rays depend on
Filament current or milliamperes

Voltage applied or kV controls
Penetration

Tube current or mA controls
Exposure time

3.1.8 Quality and Quantity of X-Rays

The quality or the energy of X-rays produced depends upon the energy of the electrons striking the target. The energy of the striking electrons is controlled by the accelerating potential which is of the order of kilovolts (kV). An increase in kV results in the generation of more energetic X-rays (Fig. 3.1.6) i.e. X-rays of shorter wavelength.

The quantity or the intensity of X-rays produced depends on the number of electrons hitting the target i.e. tube current. The tube current, which is of the order of milliamperes, can be controlled by controlling the number of electrons emitted by the filament i.e. by controlling the heating current through the filament. Increasing the milliamperage increases the number of electrons. This in turn increases the quantity or the intensity of X-rays (Fig.3.1.7) which helps in reducing the exposure times.

In general, if the intensity of X-radiation from an X-ray tube is plotted graphically as a function of photon energy, then the output spectrum comprises of a continuous spectrum superimposed on which is the characteristic X-ray spectrum. The continuous X-ray spectrum also has a short wavelength limit.

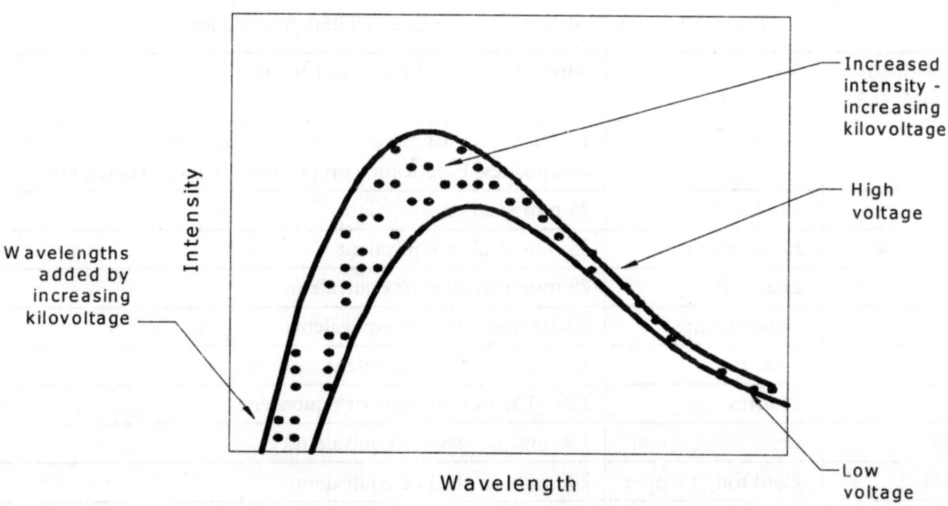

Fig 3.1.6 Effect of kV on the X-Ray Spectrum

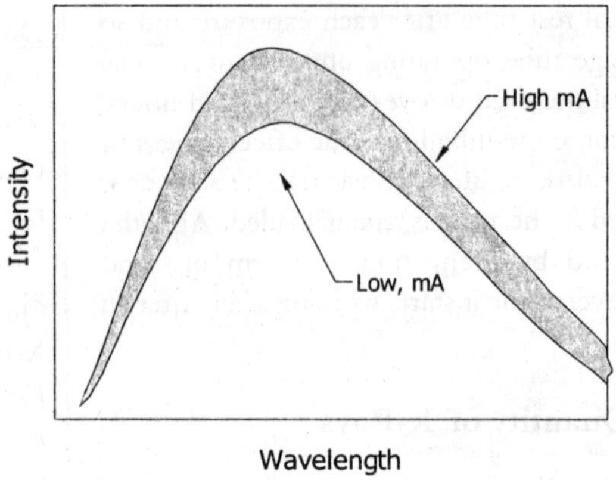

Fig. 3.1.7 Effect of mA on the X-ray spectrum

3.1.9 Modern X-Ray Equipment

A lot of development has occurred towards the production of versatile X-ray equipment. A wide range of sets are now commercially available which are compact, portable and with very stable and increased X-ray outputs. Most of these units are microprocessor controlled with auto warmup and programming facility for storing the exposure parameters. Conventional knobs have been replaced by touch screen sensors for varying the kV, mA or time.

X-ray machines can be classified by the size of their focal spots as conventional (1 mm and more) mini focus (0.1 mm – 1 mm) or microfocus (< 0.1 mm). Another rough method of classification is according to their maximum voltage as conventional (0-450 kV) or high energy (> 1 MeV). The choice among the various classes will depend on the type of work to be done. Table – 3.1.1 lists the voltage ranges and provides a rough guide of the approximate thickness that can be examined.

Table 3.1.1 Typical X-ray Machines and Their Applications

Maximum voltage (kV)	Screens	Approximate Thickness Limits
50	None	1 – 1.5 mm of steel or equivalent. Upto 8 – 10 mm of composites, Electronic components; wood, plastics, etc.
150	Lead foil	25 mm steel or equivalent.
	Fluorescent	35 mm steel or equivalent
300	Lead foil	75 mm inch steel or equivalent.
	Fluorescent	90-100 mm steel or equivalent.
420	Lead foil	110 mm steel or equivalent.
	Fluorescent	120 –130 mm of steel or equivalent.
1000	Lead foil/Copper	150 mm of steel or equivalent.
2000	Lead foil/ Copper	200 mm of steel or equivalent.
8 to 25 MeV	Lead foil/ Copper	300 – 400 mm of steel or equivalent.

Chapter 3.2
Gamma Rays

Gamma rays are electromagnetic radiations similar to X-rays but shorter in wavelength. In fact the only difference between gamma rays and X-rays is their origin. While gamma rays are produced by transitions within the nucleus, X-rays are produced outside the nucleus. Radiography with gamma rays has the advantages of simplicity and compactness of radiation source and independence from external power supply. This makes it ideal for field applications and also facilitates the examination of pipe, pressure vessels and other assemblies in which access to interior is restricted. The great penetrating power of gamma rays from most of the radioactive sources make them valuable especially when large thicknesses or high density specimens are involved. Most of the concepts relating to radioisotopes, such as, radioactivity, radioactive decay, half life, specific activity of radioisotopes, the type of radiation emitted, the interaction and consequent absorption of these radiations by matter and the general properties of these radiations, have already been discussed Section-2. In this chapter, the radioisotopes used in gamma radiography, their characteristics and handling are discussed.

3.2.1 Production of Gamma Ray Sources

Gamma rays are produced by the decay of naturally occurring or artificially produced radioisotopes. Prior to the availability of artificially produced radio isotopic sources, radium was widely used for gamma radiography. Its great merit was its very long half life of 1590 years and its large radiation output. With the advent of nuclear reactors that could generate large amounts of neutrons, artificial radioactive isotopes became available easily and economically and radium has been completely replaced by less expensive artificial radioisotopes.

Radioisotopic sources are produced mostly by activation process. When a target element is bombarded with neutrons in a reactor, activation may occur mainly by (n, γ), (n, β) or (n, α) reactions depending on the energy of the neutrons. The (n, α) reaction is mostly employed for the production of isotopic sources used in industrial radiography. This reaction is primarily a thermal neutron reaction. The neutron is captured in the nucleus of the bombarded element and the material produced is a radioactive isotope of the original element. For example,

$$_{27}Co^{59} + {}_{0}n^{1} \rightarrow {}_{27}Co^{60} + \gamma \quad ... (3.2.1)$$

Iridium -192, Thulium -170 and Ytterbium -169 are also produced in this way. Caesium 137 is produced as a result of fission of uranium 235 in a reactor.

Gamma rays
* *Electromagnetic radiations of very short wavelength*
* *Produced by nuclear transitions*

Naturally occurring/ artificially produced radioisotopes emit gamma rays

Artificial isotopes produced in nuclear reactor by activation process

Apart from nuclear reactors, particle accelerators such as linear accelerators (Linacs) are also used for the production of some radioisotopes.

3.2.2 Choice of Gamma Ray Source

While a variety of radioisotopes are produced in a nuclear reactor, only a select few have been utilised for the purposes of radiography. The rest of the other isotopes produced have been found to be unsuitable for a variety of reasons. Some of the desirable characteristics of a gamma ray source include

(a) It should have a sufficiently long half life.
(b) It should have a high specific activity. This means a small source (less unsharpness) and high gamma ray output (less exposure time).
(c) It should be easily available and preferably be of low cost.
(d) The energy of gamma rays from the source must be sufficient to penetrate the thickness of the specimen to be radiographed.

3.2.3 Sources for Industrial Radiography

The four most popular radiographic sources are : Cobalt 60 (Co^{60}), Caesium 137 (Cs^{137}), Thulium 170 (Tm^{170}) and Iridium 192 (Ir^{192}).

3.2.3.1 Cobalt-60

Cobalt occurs as a single isotope Co-59 which is transformed into Co-60 after neutron capture in a nuclear reactor. The isotope decays with a half life of 5.3 years. It emits two gamma rays of energy 1.17 MeV and 1.33 MeV. The penetrating power is comparable to a 2-3 MeV X-ray generator. It is used for the radiography of iron, copper and other medium weight metals ranging in thickness from 25 mm – 225 mm. It can also be used to radiograph thinner sections of heavy metals such as tantalum or uranium.

3.2.3.2 Iridium-192

Ir 191 on activation by the (n,γ) reaction in a nuclear reactor produces Ir 192. This source has a half life of 74.3 days. A number of gamma rays (~24) are produced during the decay process. The average energy of the emitted gamma rays is 0.45 MeV. The penetrating power is comparable to a 600 kVp X-ray machine. This is principally used for the radiography of steel upto 75 mm. Its relatively low energy and high specific activity combine to make it an easily shielded source with a small source size (focal spot). This makes the source an ideal choice for field work, where portability and compactness is a necessity.

Choice γ ray source depends on
* *half life (HL)*
* *specific activity*
* *Energy of radiation*
* *Cost*

Widely used sources
* *Ir 192, Co 60, Cs 137, Tm 170*

Cobalt 60
* *HL : 5.3 yrs*
* *Energy : 1.17 & 1.33 MeV*
* *Maximum Thickness (MTh): 225 mm of steel*

Iridium 192
* *HL : 74.3 days*
* *Energy : 0.45MeV (avg)*
* *MTh : 75 mm of steel*

Caesium 137
* *HL : 30.1 years*
* *Energy : 0.661MeV*
* *MTh : 100mm of steel*

3.2.3.3 Caesium -137

This is a by product of the fission process and has a half life of 30.1 years. It emits a 0.662 MeV gamma ray equivalent in energy to a 1 MeV X-ray machine. It is mainly used for the radiography of steel of thickness between 40-100mm. Cs-137 is usually handled in the form of Caesium Chloride CsCl.

3.2.3.4 Thulium-170

Thulium-170 is obtained by neutron activation of Thulium 169. It decays with a half life of 129 days. During the decay process, 0.084 MeV and 0.052 MeV gamma rays are also emitted. This isotope has been used for radiography of the objects with thicknesses as low as 0.8mm of steel or 13mm of Al with a 2% sensitivity. Sources with sizes of the order of 0.1mm can be produced. However the isotopes can be produced in limited activity only.

3.2.3.5 Other Gamma Ray Sources

Apart from the above, Ytterbium 169 and Selenium 75 have also been used for radiographic applications. Yb-169 decays with a half life of 31 days emitting a number of gamma rays with energy in the range of 0.063 MeV – 0.309 MeV. This source has been used for the examination of small diameter (about 250 mm) pipes. Selenium-75 is produced from Selenium-74 by the (n,γ) reaction by irradiation in a nuclear reactor. Selenium-75 decays with a half life of 121 days. There are several low energy gamma ray lines in the spectrum. The source has been used for the examination of thin specimens. However use of this radioisotope has not been much reported. Some of the radiation characteristics of the isotopes are summarised in Table 3.2.1 below.

3.2.4 Equipment Design

As gamma ray sources emit radiation at all times and in all directions, it is not safe to use them directly in the capsule form for radiography. They are handled in specialised equipment capable of housing the source so as to give a controlled radiation beam as desired. Such an exposure device is called "Radiographic Camera" or "Gamma Camera". The design of gamma radiography camera has progressed over the years with the object of ensuring safe operating conditions for radiographers on site. Early models of containers used a simple lead casting for housing the radioactive source and give radiation protection. The source was manipulated by a long pole to position it over the object. The later models used containers with movable shutters where the source could be stored safely. By removing a shielding component, a beam of radiation would emerge from the beam port.

The more recent development is the use of remote control to wind the source from the shielded position to the exposure point. This is achieved using a flexible source holder with a disconnectable link to a manual wind out cable. The source is moved through an

> Thulium 170
> * *HL : 129 days*
> * *Energy :0.052 MeV & 0.084 MeV*
> * *MTh : 10 mm of steel*
>
> Other gamma ray sources
> * *Ytterbium 169*
> * *Selenium 75*
> * *Gamma ray sources handled using radiographic cameras*
>
> Present day cameras
> * *Made of depleted uranium*
> * *Are of remote controlled type*

armoured flexible source guide tube, which has an end stop, which is normally fitted into a beam limiting collimator at the weld position. The projector type container has the advantage of remote operation, which gives additional safety by positioning the operator at a considerable distance away from the source throughout the exposure. The present containers mostly use depleted uranium for radiation shield. This shielding is either in the form of a compact spherical casting with a movable shutter which allows the source to be projected or alternatively uses a larger casting with an 'S' tube internal path. The "S" tube design gives full radiation protection with the advantage of no moving parts in the primary shield (Fig. 3.2.1).

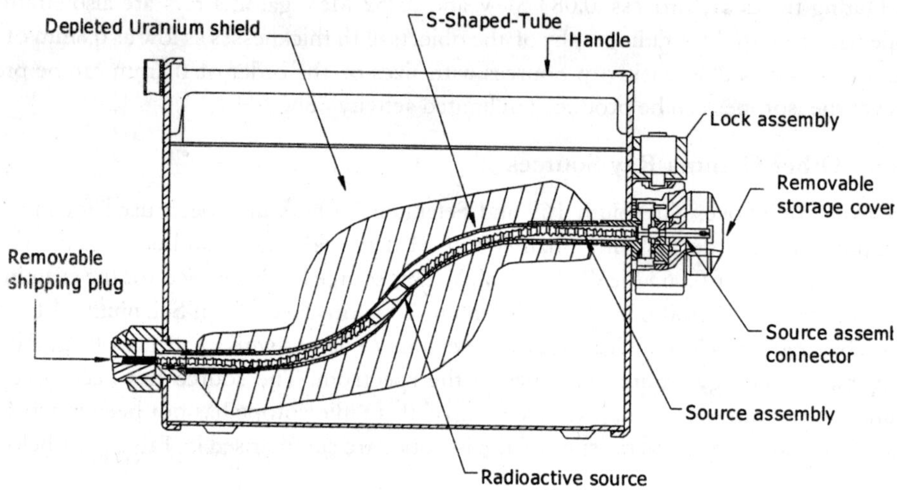

Fig 3.2.1 Typical remote controlled radioisotope camera with "S" tube design.

3.2.5 Quality and Quantity of Gamma Rays

In contrast to X-ray machines, which emit a broad and continuous band of wavelengths, gamma-ray sources emit one or a few discrete wavelengths typically as shown in Fig. 3.2.2 below.

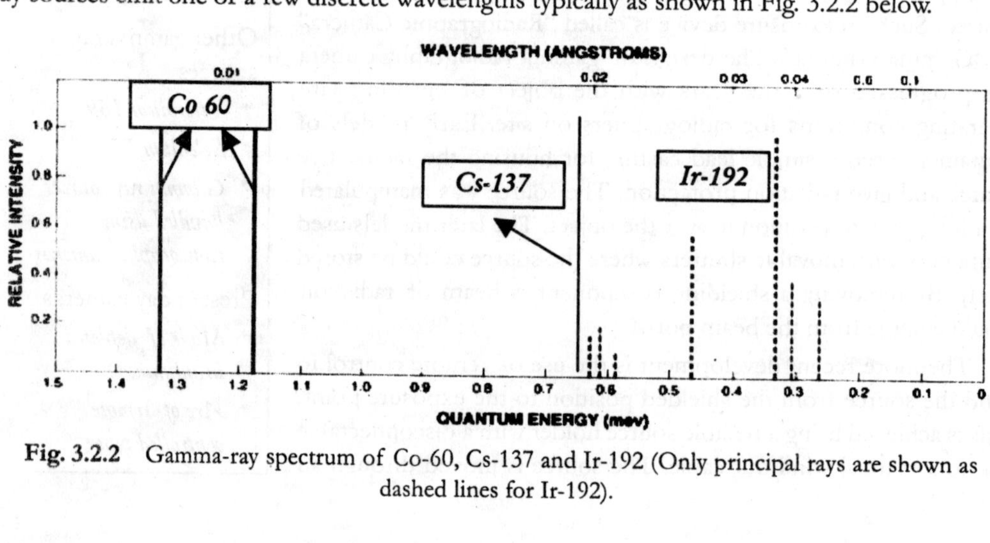

Fig. 3.2.2 Gamma-ray spectrum of Co-60, Cs-137 and Ir-192 (Only principal rays are shown as dashed lines for Ir-192).

In industrial radiography, gamma rays are specified in terms of the energy of the individual photon. The unit of energy used is the electron volt (eV). The quantity or *intensity* of gamma radiation depends on the strength of the radioisotopic source used—specifically, on the number of radioactive atoms in the source that disintegrate in one second. This is given in terms of Curies. The energies of radiation emitted by a gamma-ray source, and their relative intensities, depend *only* on the nature of the radioisotope. Thus radiation *quality* and *quantity* of emitted by a given radioisotopic source cannot be varied at the will of the operator. Gamma-ray sources gradually lose activity with time, the rate of decrease of activity depending on the half life of the source. This decrease in activity necessitates more or less frequent revision of exposures and periodic replacement of sources. Fig. 3.2.3 depicts the decay curve for Ir-192. Both linear and logarithmic plot are shown.

Gamma ray spectrum is discrete

Intensity of gamma rays emitted depends on source strength

Quality and quantity of gamma ray emitted cannot be varied by operator.

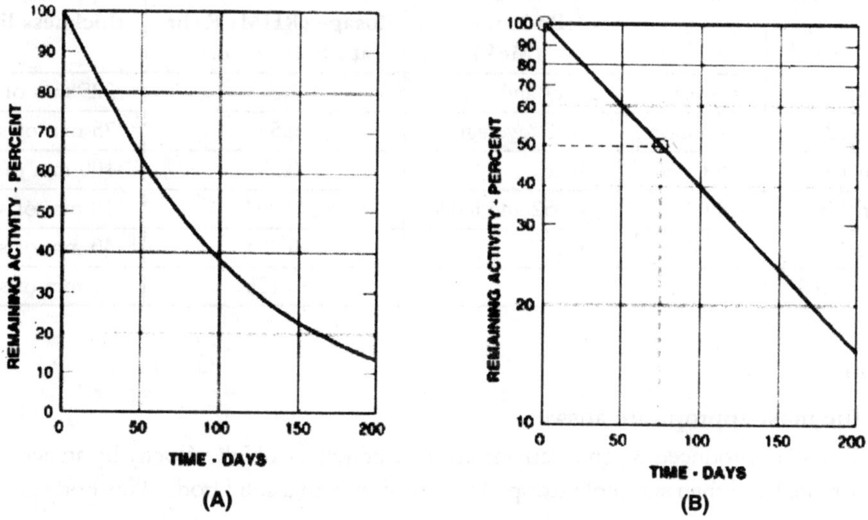

Fig. 3.2.3 Typical decay curve for Ir – 192. (a) is a linear plot (b) is a logarithmic plot.

3.2.6 Advantages and Limitations of Gamma Ray Sources

There are a number of advantages in using a gamma ray source for radiography. Gamma ray equipment do not require electrical power for operation and therefore are ideally suited for field applications and remote areas. Being compact and considerably lighter in weight compared to X-ray equipment, they are easily transportable. Gamma ray equipments need minimum maintenance unlike X-ray units. Since the isotopic source material is small, it facilitates examination of pipes, pressure vessels and other assemblies in which the access to interior is restricted.

The limitations of gamma ray sources include

(a) The output of a source is limited. If the output is to be increased without increasing the physical size of the source, the specific activity is to be increased. The specific activity is controlled by the neutron flux available in the irradiation positions in a reactor. In the case of X-ray, the output can be increased by increasing the current.

(b) The energy of radiation cannot be varied to suit the thickness of the specimen as in the case of X-ray units wherein the energy can be varied over a range. This results in radiographs with poor contrast and sensitivity especially during thin section radiography.

(c) Gamma ray sources decay with time. This necessitates increasing exposure period as the source ages and periodic replacement of weak sources

(d) Since isotopic sources are continuously emitting, unlike X-rays machines which can be turned off they should be stored safely with adequate precautions.

Table 3.2.1 Characteristics of Commonly used Industrial Radioisotopic Sources

Isotopic Source	Half Life	Energy of Radiations (MeV)	Gamma ray dosage (RHM) R/hr at 1 m for 1 Curie	Approximate thickness limit
Cobalt 60	5.3 years	1.17 and 1.33	1.3	225 mm of steel
Iridium 192	74.3 days	0.45 (average)	0.5	75 mm of steel
Caesium 137	30.1 years	0.661	0.39	100 mm of steel
Thulium 170	129 days	0.052 and 0.084	0.003	10 mm of steel
Selenium 75	120 days	0.217	0.203	30 mm of steel
Ytterbium 169	32 days	0.145	0.125	10 mm of steel

Questions

Choose the most appropriate answer

1. X-rays are produced when electrons are accelerated to a high velocity by an acceleration potential and then suddenly stopped by collision with a solid body. This body is called

   a. cathode　　b. filament
   c. target　　d. generator
   e. copper block

2. The efficiency of X-ray production is best when the target material has a:

   a. low atomic number　　b. high atomic number
   c. low hardness　　d. high hardness
   e. high melting point

3. The device that is used to convert alternating current from the high voltage transformer to direct current is

   a. rectifier　　b. cathode ray tubes
   c. alternator　　d. vacuum x-ray tube
   e. autotransformer

4. Two X-ray machines operating at the same kilo voltage and having identical mA settings:
   a. will produce the same intensities and qualities of radiation
   b. will produce the same intensities but may produce different qualities of radiation
   c. will produce the same qualities but may produce different intensities of radiation
   d. may give not only different intensities but also different qualities of radiation
   e. none of the above
5. The ability of an X-ray beam to penetrate a given thickness of material is determined by:
   a. kilovoltage
   b. time
   c. milliamperage
   d. source-to-film distance
   e. Curie strength
6. Co-60 used in nondestructive testing emits:
   a. alpha particles
   b. neutrons
   c. Gamma rays
   d. ionising radiations
   e. X-rays
7. Short wavelength electromagnetic radiation produced during the disintegration of nuclei of radioactive substances are called:
   a. rayleigh rays
   b. X-rays
   c. Gamma rays
   d. ultraviolet rays
   e. electromagnetic radiation
8. As the accelerating potential applied to the X-ray tube is increased:
   a. X-rays of longer wavelength and more penetrating power are produced
   b. X-rays of shorter wavelength and more penetrating power are produced
   c. X-rays of shorter wavelength and less penetrating power are produced
   d. X-rays of longer wavelength and less penetrating power are produced
9. In order to decrease the exposure time during radiography:
   a. the acceleration potential should be increased
   b. the tube current should be increased
   c. the tube current should be decreased
   d. the test specimen should be moved farther from the film
   e. a lower kilovoltage should be applied to the tube
10. A Co-60 source has a half-life of:
    a. 530 months
    b. 361 months
    c. 61.2 months
    d. 74.3 days
    e. 129 days
11. The voltage and waveform applied to the X-ray tube by a high voltage transformer primarily determines the:
    a. quantity of radiation
    b. duration of exposure
    c. penetrating ability
    d. X-ray beam divergence
    e. quality of exposure

12. The small area in the anode of an X-ray tube from which the radiation emanates is called the:
    a. target
    b. focal spot
    c. focusing cup
    d. cathode e. anode
13. The quality of radiation emitted by a radioisotopic source :
    a. depends on the focal spot/diameter of isotopic source
    b. depends on the nature of the isotope used.
    c. Can be varied by the operator
    d. greater in Ir-192 than in Co-60
    e. depends on half life of the isotope
14. All the sources except one is produced by neutron capture in a reactor
    a. Ir 192
    b. Th 170
    c. Co 60
    d. Caesium 137
    e. None of the above
15. The most widely used reaction for the production of isotopes using nuclear reactor is
    a. (n, μ) reaction
    b. (n, γ) reaction
    c. (n, α) reaction
    d. (n, α) reaction
    e. All of the above

SECTION 4

PHOTOGRAPHIC RECORDING

After reading this section, you should be able to answer

- ★ *How the cross – section of a film looks like and explain the functions of various constituents of the film.*
- ★ *What are the important physical characteristics of the film?*
- ★ *How to process the films.*
- ★ *The functions of the various processing chemicals used.*
- ★ *How to store exposed and unexposed films.*
- ★ *What is the care that should be taken while processing films and their consequences.*

Chapter 4.1
Characteristics of X-Ray Film

When an X-ray beam reaches the object, it contains no useful information. After the beam passes through and interacts with the object being examined, it contains all the information that can be revealed by that particular radiographic examination. This is represented by variation in the number of X-ray photons in different areas of the emergent beam. This information must be transferred to a medium suitable for viewing by the eye. The most widely used material to decode the information carried by the transmitted X-ray beam is the film. The type of film used in industrial radiography plays an important role in the detectability of defects. Hence any practicing radiography professional should have adequate knowledge about the construction and characteristics of films. This is dealt with in the following sections.

4.1.1 Film

X-ray film consists of a photographically active, or radiation-sensitive emulsion that is usually coated on both sides of a transparent sheet of plastic, called the base. Firm attachment between the emulsion layer and the film base is achieved by use of a thin layer of adhesive. The delicate emulsion is protected from mechanical damage by layers known as the supercoating. Fig. 4.1.1 shows the typical cross-section of an industrial X-ray film.

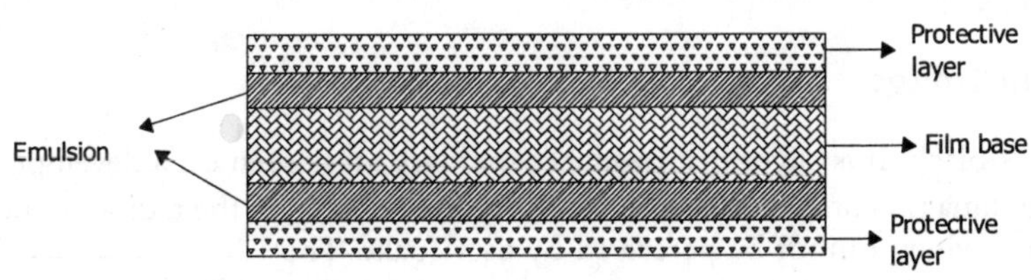

Fig. 4.1.1 Typical cross-section of an industrial X-ray film

> X-ray film consists of a transparent base with light sensitive emulsion coated on both sides.
> The film base is a sheet of transparent plastic that provides support to the emulsion.
> Polyester is the most widely used base material.

4.1.1.1 Film Base

The function of the film base is to provide a support for the fragile photographic emulsion. A film base should have three essential characteristics namely:

1. It must not absorb too much light when the radiograph is viewed.
2. The flexibility, thickness and strength of the base must allow for case of processing (developing) and produce a radiograph that "feels right" when handled.

3. The base must have dimensional stability; i.e. the shape and size of the base must not change during processing or during stored life of the film.

While original X-ray plates consisted of glass plates, which were replaced by cellulose nitrate in 1914, present day films use polyester as the base material. Polyester as a film base offers the advantage of improved dimensional stability, even when stored under conditions of varying humidity, and it is much stronger than cellulose nitrate or acetate. Present day bases are tinted blue so that the film is easier to look at with minimal eyestrain.

4.1.1.2 Emulsion

The two most important ingredients of a photographic emulsion are gelatin and silver halide. Gelatin helps in keeping the silver halide crystals dispersed and gives permanence to them. Silver halide crystals are the light sensitive portion of the film and are well dispersed throughout the emulsion. The silver halide is primarily silver bromide with a small amount of silver iodide. The exact composition of the various emulsions is a closely guarded industrial secret. Most X-ray films are made for use with intensifying screens, and has emulsion coated on both sides of the base. Emulsion thickness varies with film type, but is usually not thicker than 0.5 mm. A thicker emulsion would not be useful because of the inability of light to penetrate the deeper layers.

Photographic emulsion would not adhere to the finished base if applied directly. Therefore, a thin layer of adhesive substance is applied to the base to ensure perfect union between base and emulsion.

4.1.1.3 Protective Coating

Covering the emulsion is a thin layer, commonly gelatin that serves to protect the emulsion from scratches, abrasion and mechanical damage.

4.1.2 Latent Image

Metallic silver is black. It is silver that produces the dark areas seen on a developed radiograph. When X-, or gamma rays or electrons interact with the emulsion of the radiographic film, a latent image is formed which cannot be perceived by us visually. The basic concept of the formation of latent image was formulated by Gurney and Motley. According to their hypothesis, when an atom in the silver bromide crystal absorbs energy from a ionising particle such as X- or gamma ray photon, an electron is liberated. The electron can move in the crystal for relatively large distances as long as it does not encounter a region of impurity or fault in the crystal. A site of crystal imperfection, such as a dislocation defect, or an AgS sensitivity speck, may act as an electron trap where the electron is captured and temporarily fixed. The electron gives the sensitivity speck a negative charge. This attracts the mobile interstitial Ag+ ions in the crystal. This union creates an atom of metallic silver at the sensitivity speck. The single atom of silver then acts as an electron trap for a second electron. The

> Emulsion consists of light sensitive silver halide crystals well dispersed in gelatin medium.
>
> The sensitivity speck provides the site for the formation of the latent image
>
> The invisible deposition of metallic silver at the sensitivity speck is the latent image.

negative charge causes a second silver ion to migrate to the trap to form two-atom silver nucleus. Thus, growth of silver atoms at the site of the original sensitivity speck continues by repeated trapping of electrons, followed by their neutralization with interstitial silver ions. This latent image distinguishes the silver bromide grain from the other "unexposed" grains making it susceptible to attack by the developer. More the radiation interacting with the silver halide, more metallic silver deposition would occur. If more silver is deposited in the emulsion, it means greater radiographic density on the film.

4.1.3 Film Properties

The accuracy of a radiographic film examination depends, in part, on the visibility of important information on the film. Understanding the relationship between the exposure a film receives and the way the film responds to the exposure is essential for intelligent selection of proper exposure factors and type of film to provide maximum information content in the radiograph.

What is meant by the term "exposure" of an X-ray film or film-screen combination? Exposure is the product of intensity and time. In case of X-rays, this exposure is the product of the milliamperes of X-ray tube current (at a given kV) and time while for gamma ray sources; it is the source strength in Curies and the exposure time. Thus, an exposure of 10 milliamperes for 1 minute is expressed as 10 milliampere-minutes, usually written 10 mA-mins. An exposure of 10 mA-min could also be produced by using 5 mA tube current for 2 minutes, 20 mA for 0.5 min, and so forth. Exposure of the X-ray film produces film blackening, or density. Two general but not completely accurate statements should be kept in mind:

1. The intensity of the radiation beam (milliampere-minutes or Curie - hours) controls the film density
2. Penetration or kV/keV controls the image contrast

4.1.4 Radiographic Density

Radiographic density is the black metallic silver deposited in the emulsion of X-ray film after exposure to light or radiation after processing. It is a quantitative measure of film blackening and is defined as the logarithm of the ratio of the incident light striking the radiographic film to the light transmitted through the film. Mathematically,

$$D = \log\left(\frac{I_o}{I_t}\right) \quad \text{... (4.1.1)}$$

where D is the density, I_o the light incident on the film and I_t is the light transmitted by the film. Some additional terms are also used such as transmittance or opacity. Transmittance is the ratio I_t/I_o or the fraction of the light transmitted through the film while opacity is the inverse of transmittance I_o/I_t. Taking a simple example, if 100 photons of light strike the back of the film (I_o = 100), and only one photon passes through the film (I_t = 1), then

> Exposure = Intensity × time.
>
> E = mA × time (X-rays)
>
> E= Ci × time (gamma rays)
>
> Density is the logarithm of the ratio of the incident light striking the radiograph to the light transmitted through the radiograph.

Transmittance = 1/100 = 0.01, Opacity = 100/1=100 and Density = log (100/1) = 2

Table 4.1.1 lists some common values for opacity [(I_o / I_t)], density [log (I_o / I_t)], and the light transmission. Higher density means a blacker film (less light transmission).

Table 4.1.1 Relation between Density, Transmittance and Opacity

Density [Log (I_o / I_t)]	Transmittance (I_t / I_o)	Opacity (I_o / I_t)
0	1.00	1
0.3	0.5	2
0.6	0.25	4
1.0	0.1	10
2.0	0.01	100
3.0	0.001	1000

Why is density expressed as a logarithm? Two primary reasons are:

1. Logarithms conveniently express large differences in numbers on a small scale.
2. The physiologic response of the eye to differences in light intensities is logarithmic

The primary factor that affects radiographic density is Exposure. Other factors which influence density include:

(a) type of film
(b) energy of radiation
(c) scattered radiation
(d) processing conditions
(e) collimation and filtration

Useful densities in industrial radiography range from about 1.5 to 4.0.

4.1.5 Measurement of Density

The density on a radiograph can be measured either by comparison with a film strip (density wedge) or by the use of instruments known as densitometers. The density wedge consists of a range of known densities and the density of a radiograph is measured by viewing it along side the radiograph under similar viewing conditions. This method is the least accurate of the available methods, since it is based upon a visual comparison of two pieces of film. However, a calibrated film strip serves as a useful guide to density in many instances when a densitometer is not available.

> Density is affected by type of film, energy of radiation, amount of exposure and the processing conditions
>
> Density is measured using an instrument called densitometer or by comparison with density wedge.

Densitometers (fig 4.1.2) are the most widely used instruments for measurement of density. Based on mostly photoelectric principles these devices contain a light source, lenses and photocell or other light measuring device. The light

Fig. 4.1.2 Typical densitometer used in industrial radiography
(Courtesy: M/s Pulse echo systems, Mumbai, India)

transmitted by the radiograph is collected by the photocell which gives an electrical signal proportional to the number of photons received. The output of the photocell energises an analog or digital display which provides a numerical density reading. Apart from accuracy, an important property of densitometer is its reliability. The densitometer should be capable of reproducing readings without significant variations.

4.1.6 Characteristic Curve

It is necessary to understand the relationship between the exposure a film receives and the density produced by the exposure. The relationship between exposure and density is plotted as a curve, known as the "characteristic curve" or "H and D curve" (named after F.Hurter and V.C.Driffield, who first published such a curve in England in 1890).

Characteristic curves are derived by giving the film, a series of exposures, developing the film, and plotting the resulting density against the logarithm of relative exposure. Typical characteristic curve of X-ray films is shown in Fig 4.1.3.

The characteristic curve has three portions: the toe, straight line portion and shoulder. The toe of the curve does not begin at zero since density caused by fog occurs even without exposures. The straight line portion represents the area of film response where increases in

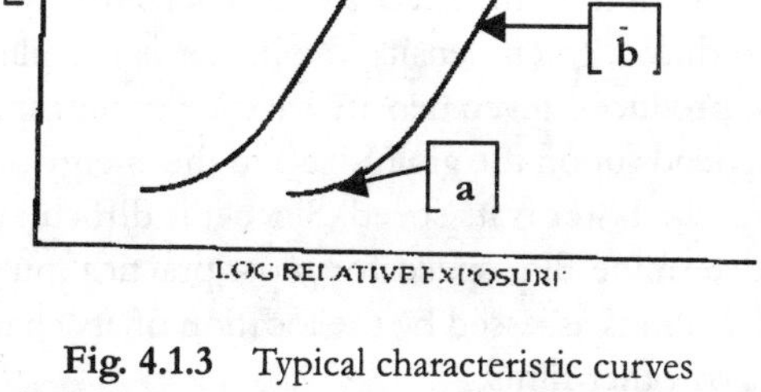

Fig. 4.1.3 Typical characteristic curves
The curve has three portions (a) toe (b) straight line portion and (c) shoulder

exposure produce increases in the density of the film. The shoulder of the curve represents the region of saturation wherein, even large increases in exposure do not produce any appreciable change in density.

4.1.6.1 Use of the Characteristic Curve

The characteristic curve is the most useful tool for a radiographer. Examination of the characteristic curve reveals much about the properties of a given film.

(i) When comparing the speed of films, the location of the characteristic curve provides a clue to which is a faster film. In fig. 4.1.3 above, film A is a faster film.

(ii) The slope of the characteristic curve gives the film contrast. High contrast films have a steeper straight line portion compared to low contrast films.

(iii) Using these curves one can calculate the exposures needed when density / films are varied.

> Fog density is due to the inherent density of the film base material and the development of some grains of silver halide even without exposure to radiation.
>
> Film speed is the sensitivity of a film to exposure and depends on
>
> - *grain size*
> - *energy of radiation*
>
> Coarse grain films are faster compared to fine grain films.

4.1.7 Fog Density

If an unexposed X-ray film is processed, it will indicate a density of about 0.12. This is called the fog density or fog level of the film. The fog density consists of base density and chemical fog density. The plastic material used to make the film base absorbs a small amount of light. Also, the blue dye used to colour film bases adds slightly to base density. Both these factors contribute to the base density. Total base density will average about 0.07. The chemical fog density is due to the fact that some grains are capable of being developed even without exposure. The actual fog density varies with the type, age and storage conditions of the film.

4.1.8 Film Speed

Generally film speed is termed as the sensitivity of film to exposure. In absolute units, film speed is defined as the reciprocal of the total dose in roentgen of a particular radiation spectrum that produces a given density on the film. Film speed is thus inversely related to exposure. The lower the exposure required to produce a given density, the higher is the film speed. Conversely films requiring large exposures to produce a given density have lower film speeds. The speed of a film, for normal development, depends upon the grain size and the energy of radiation. In general, the coarser the grain of the film, the faster is its speed. Since it is difficult and cumbersome to make absolute measurement to determine film speed for most practical purposes, relative speeds are used. The relative speed of a film is assessed by the location of its characteristic curve along the log E axis relative to curves of other films.

In Fig. 4.1.4 the characteristic curves for various radiographic films are spaced along the log relative exposure axis. The spacing of the curve arises from the difference in relative speed - the

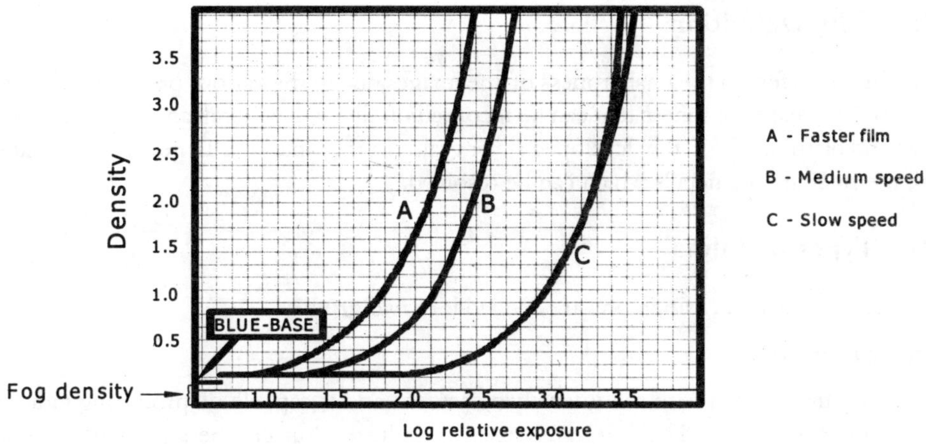

Fig 4.1.4 Characteristics curves of X-ray films of different speed

curves for the faster films lying towards the left and those for the slower films towards the right. From these curves relative exposure to produce a fixed density can be read; the relative speeds are inversely proportional to these exposures.

The amount of exposure of a given radiation energy required on a film to produce a given film density (usually 2.0) under specified conditions of processing is called 'film factor'. A film with lower film factor is faster than the one with a higher value. Film factor depends on energy of radiation.

The slope of the characteristic curve is the contrast.

Average gradient is the slope of the straight line between two density points on a characteristic curve.

Fine grain films have better film definition compared to coarse grains.

4.1.9 Film Contrast

Film contrast or gradient is defined as the change in density recorded on a film for a given change in radiation intensity. The slope of the characteristic curve at a given density is the measure of the film gradient, G_d at that density. High contrast films have steep straight line portion compared to low contrast films. Numerically film contrast is expressed through average gradient. The average gradient is defined as the slope of straight line joining two density points on the characteristic curve. Mathematically it can be expressed as:

$$G_{d\,(avg)} = \frac{[\text{Density}_2 - \text{Density}_1]}{\text{Log. Rel. Exposure}_2 - \text{Log. Rel. Exposure}_1} \qquad \text{... (4.1.2)}$$

The gradient of a film depends on the size of its crystals. It is practically independent of the radiation energy. Gradient of industrial X-ray films goes on increasing with optical density. In general, fine grain films have lower speeds, but, higher gradient at a given density, than coarse grain films. There is very little effect on the shape of the characteristics curve due to energy of radiation.

4.1.10 Film Definition

Film definition refers to the unsharpness of the images due to the various properties of the film. The degree of image sharpness that can be recorded on film will depend, among other things, on the size and distribution of the developed silver grains in the emulsion. In general the smaller the grains, the finer are the details which can be resolved.

4.1.11 Types of Films

Films used in industrial radiography are divided into two groups:

Salt screen type film

This type of film is used with salt intensifying screens and is capable of producing radiographs with minimum exposure. They are rarely used in industry, but can be advantageous where a reduction in exposure is important especially when large thicknesses are to be examined.

Direct type films

These are films which are intended for direct exposure to X-rays or gamma rays, or for exposure using lead screens. Some of these films may also be used with fluorometallic screens. Direct type films are predominantly used in industrial radiography.

Various companies manufacture a variety of radiographic films to meet the wide range of demand by industries. Each type of film is designed to meet certain requirements and these are dictated by the inspection conditions such as

(a) part density and thickness
(b) energy and intensity of radiation
(c) level of inspection and sensitivity desired

No single film can meet all the demands. Thus a number of different types of films are available, the choice of which is governed by the combination of the radiographic technique and the films to obtain the best desired result.

The film factors that must be considered in choosing a film are speed, contrast, latitude and graininess. All radiographic films can be grouped according to these four factors. This is given in Table 4.1.2 below.

4.1.12 Packaging of Films

Radiographic films are available in a variety of packing such as day pack, envelope packed, folder wrapped, etc. These may be in roll or sheet form. Day pack films are ready to use films. Depending on the need, day pack films are available with and without lead screens. These films are however costly. The most widely used form is the cut sheet. A pack of 50 or 100 cut sheet films is sealed in a pouch and the pouch sealed in a box with reclosable top.

Each film is wrapped on both sides with a piece of paper called the interleaving paper. In some cases, alternate films are interleave wrapped. This is called as alternate interleaving.

Table 4.1.2 Classification of Industrial X-ray Films

Film Characteristic	Grain	Speed	Contrast
Type I - For critical inspection and radiography of light metals and alloys	Very Fine	Very slow	High
Type II - For radiography of light metals and heavier steel components/parts at 1 – 2 MeV	Fine	Slow	High
Type III - Highest available speed when gamma rays or high energy X-rays are used direct or with lead screens	Medium	Medium	Medium
Type IV - Highest available speed and high contrast when used with fluorescent screens. Lower contrast when used director or with lead screens	Medium	Medium	Fast

4.1.13 Film Cassettes

Film cassettes can be flexible or rigid.

(a) Flexible cassettes are made of PVC and find extensive use in the field since they can be adapted to various shapes and sections such as circumferential welds etc. The flexible cassettes are of two types– double envelope type or single envelope type.

(b) Rigid cassettes consist of thin aluminum front with a felt pressure pad to provide intimate contact between the film and the screen.

> Intimately screens
>
> - *Reduce exposure times*
> - *Improve radiographic contrast by absorbing longer wavelength radiations.*
> - *Normally screens are used above 120kV*

4.1.14 Intensifying Screens

When an object is radiographed without using screens, there are two disturbing factors.

1. There is lesser absorption of transmitted radiation intensity in film, which results in longer exposure time.
2. Low energy scattered radiation is absorbed more, which spoils the image quality. Hence, to reduce exposure time and improve image quality of radiographs, screens are used. There are two types of screens commonly used in industrial radiography: metallic foil screens and fluorescent screens. These screens differ in their basic characteristics e.g., speed, contrast and elimination of scattered radiation.

4.1.14.1 Metallic Foil Screens

Metallic screens help in reduction of scattered radiation from the object. The X-ray film is sandwiched between a pair of screens called the front screen and back screen. Absorption of incident radiation in the object results in production of low energy scattered radiation and this falls on the film, along with the transmitted primary radiation. As high atomic number elements have more absorption capacity for low energy radiation, if a metal foil of high atomic number is placed between object and film, it will absorb most of the low energy radiation, before it falls on the film. Hence, effect of the low energy, scattered radiation will be reduced. This metallic foil will serve additional purpose of intensification action also. Similarly, to reduce low energy back scattered radiation (obtained due to back scattering of the primary radiation, from objects placed behind the specimen to be radiographed), a high atomic number sheet is placed behind the film. Hence, a double coated X-ray film is sandwiched between a pair of metallic foil screens to reduce the effect of scattered radiation on the film, thus resulting in an improved definition of radiograph. Normally the back screen is thicker than the front screen.

In addition to absorbing the scattered radiation, metallic screens also intensify the primary radiation. The screens absorb primary radiation and produce photoelectrons (through photoelectric effect) which are more efficiently absorbed by the film. Lead foil is the most commonly used intensifying screen. Lead screens are normally used only above 120 kV. Apart from lead, copper and tantalum screens have also been used. The intensifying action of a radiography screen is expressed in terms of Intensification Factor (IF) which is defined as

$$\text{IF} = \frac{\text{Exposure time required to produce a certain film density without screen}}{\text{Exposure time required to produce same density with screen}}$$

In this definition, it is assumed that the same film and radiation source are used for both the exposures. 'IF' due to double lead screen exposure with Ir–192 source is about 4.5.

Intensification factor due to metallic screens depends on:

(a) foil element,

(b) thickness of foil,

(c) energy of radiation, and

(d) specimen thickness.

4.1.14.2 Salt or Fluorescent Screens

These are also called 'salt screens'. These screens consists of a cardboard or plastic material on which certain inorganic crystalline substances, which glow when X or gamma rays fall on them, are coated on one side. These screens are used in pairs, so that a double coated X-ray film can be sandwiched between them. Intensifying action of a fluorescent screen is due to additional glow appearing as a result of interaction of X- or gamma ray with the salt screen material. A material, which has its light emission in blue or ultraviolet region, is used for this purpose, since the film is highly sensitive to blue light. The most commonly used material is calcium tungstate. During X- or gamma ray exposure, each crystal in salt screen emits light and this light diverges in all directions and gives image unsharpness. Thus salt screens produce higher screen unsharpness compared to lead screens. Because of this, these screens are not much used in industrial radiography.

Chapter 4.2

Darkroom Techniques and Film Processing

In any radiographic facility, adjacent to the film viewing room is a film handling cum processing room. It is well known that X-ray films are sensitive to light. Hence, to protect it from light during loading, unloading and processing, the film must be handled in a dark room. The darkroom is not completely dark but is darkened using appropriate accessories so that unwanted and stray light does not enter. Good planning and layout of the darkroom is essential for successful radiography. It should be designed to meet individual requirements based upon the amount and nature of the work being done. The various activities in the dark room include

(a) storage of unexposed films
(b) film loading and unloading
(c) film processing

4.2.1 Handling of X-Ray Films

Films should always be handled carefully to avoid physical strains such as pressure, creasing, buckling and friction and marks resulting from unclean or moist hands. The best way to hold the film is at the edge using fingertips. Whenever the films are loaded in flexible or rigid cassettes and external clamping devices are used, it should be ensured that the pressure is applied uniformly. While normal pressure applied on a cassette to provide good contact with the object does not cause any mark or damage on the films, localised pressure on film can produce desensitised areas resulting in marks on the films. Films should be drawn smoothly from cartons and exposure holders. Rapid movements may cause electric discharges, which appear as dark circular or tree like markings on the film (Fig. 4.2.1). Before the film is loaded between the intensifying screens, the interleaving paper should be removed.

Fig. 4.2.1 Film artifact due to static electric discharge

Good planning and layout of darkroom essential for successful radiography

X-rays films are to be handled with care to avoid film marks and artifacts

4.2.2 Film Storage

There are two different storage periods for the films – one before processing i.e. the unexposed films and after processing for archival purposes. Unexposed films are to be stored in cool dry areas free from dust, radiation, chemicals, toxic fumes etc. In case the darkroom is close to any radiation areas, the films should be stored in lead lined boxes. The thickness of the lead lining depends on the intensity of radiation and the length of time for which the films would be kept there. Films may be stored in refrigerator for long periods. It is always preferable to store the individual boxes of films on edge. The methodology of stacking should be such that the oldest films are used first since films deteriorate with age. This is the FIFO i.e. the first in first out system. It is preferable that the storage temperature be between 4°C and 24°C and the relative humidity between 30 to 60%. Periodic measurements are necessary to ensure that the temperature and humidity are within limits.

> Films are to be stored in cool dry areas free from dust, fumes, chemicals and radiation.
>
> Individual boxes to be stacked on edge with older films first and new films following. This is the principle of FIFO- first in first out.
>
> Developing - the step during which the latent image becomes visible.

4.2.3 Manual Film Processing

On exposure to radiation, a latent image of the object, invisible to the human eye is formed in the X-ray film. The purpose of film processing is to convert this invisible image to a visible and permanent image. The processing of film is an important step in the overall radiographic process and should be done with care. Film processing can be done either manually or through the use of automatic processors. Here we outline manual film processing. The major steps in manual processing are developing, rinsing, fixing, washing and drying. These are carried out in solutions contained in tanks as shown in Fig. 4.2.2.

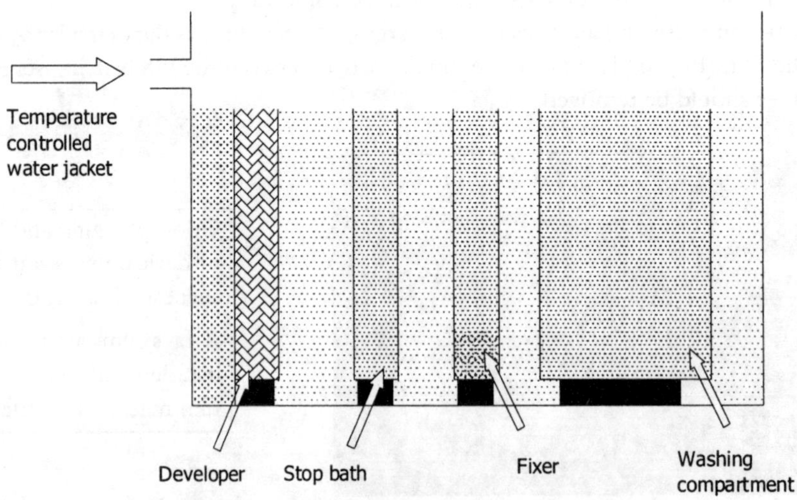

Fig. 4.2.2 Schematic of Manual Processing

4.2.3.1 Developing

Developing is the step during which the latent or invisible image becomes visible and the reduction of exposed grains of silver halide into metallic silver occurs. The developer used for processing of X-ray films essentially has five major components

(1) Developing agent – commonly used agents are metol, hydroquinone and phenidone. Developing agent is a reducer and its function is to reduce the silver halide to metallic silver by donating an electron.

(2) Accelerators – sodium or potassium carbonates are normally used. They serve to increase the activity of the developer.

(3) Preservative – usually sodium sulphite or sodium sulphate. It prevents the developer from being oxidised by air.

(4) Restrainer – usually potassium bromide or potassium iodide, it controls the activity of the developer and retards the development of unexposed silver halide crystals thus controlling overall fog.

(5) Solvent – The solvent for the developer is water which dissolves and ionises the developer chemicals. Water also aids in softening the emulsion.

Apart from the above main constituents, the developer also has varied quantities of other ingredients that are added for special purposes to improve the physical property, activity, life and reduce sludging.

Practical points to be considered before processing

- **Check level of developer, stop bath and fixer in the tanks. Level should be adequate to cover the cross bars of film hangers. In case level is low, replenish the same.**
- **Check temperature of solutions. It is preferable to have temperature around 20° C.**
- **Stir all solutions**
- **Ensure a steady and sufficient flow of water in the washing tank**
- **Clean all working surfaces and wash hands thoroughly.**
- **Check safelights are in working condition and filters are not cracked.**

4.2.3.2 Time Temperature Relationship

Two factors namely temperature and time control the rate of development of the film. These two factors have an inverse relationship i.e. as the temperature of the developer increases, the time of development decreases and vice versa. This is a very important relationship that controls the actual development and hence density achieved on the radiograph. All manufacturers of solutions provide time – temperature relationship data with their chemicals. A typical time – temperature chart is given in Fig. 4.2.3. The best results are obtained at a temperature of 20°C. A range of about 18 – 24°C would produce satisfactory radiographs with correct development times. Higher developer temperatures would result in chemical fog and also increased graininess. Also the developer deteriorates faster. At high temperatures, the emulsion of the film may undergo reticulation and get detached. At temperatures lower than 18°C, the development times become high and the contrast producing agent in the developer gets restrained leading to poor quality radiographs.

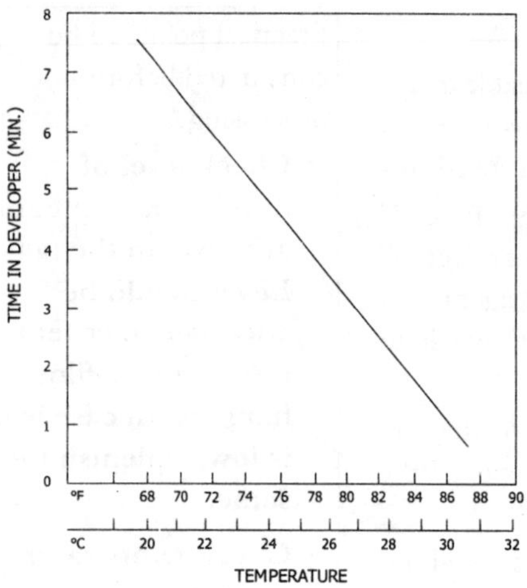

Fig. 4.2.3 Typical Time – Temperature Chart

Hints to Practical Developing Procedure

- **Check all doors are closed. Turn off all lights and proceed to work under safelight conditions only.**
- **Check level of Ensure that developer solution has been stirred well. Solutions tend to separate out when left undisturbed.**
- **Check level of Unload cassette and gently remove the film from the cassette and screen.**
- **Hold film at the edge and load it in the hanger. Ensure that the film is gripped well with uniform spacing all around the hangar.**
- **Immerse film steadily into the developing solution**
- **Set timer after ascertaining developer temperature and estimating the time from time temperature chart**
- **Tap the film hangar to dislodge any air bubbles that might be formed on the surface of the film.**
- **Agitate the film for ~ 15 seconds initially and for ~10- 15 sec. after each minute of development.**
- **At the end of the developing time, remove film from the tank, drain back excess developer for approximately 2 to 3 seconds only**

4.2.3.3 Replenishment

As a developer is used, its developing power decreases due to

(i) developing agents and preservatives getting consumed,

(ii) accumulation of reaction products such as hydrogen ions and bromide

(iii) oxidation of developer by atmosphere.

To maintain the chemical activity of the solution so that best and consistent results are obtained and also to maintain the level of the solution in the tanks so that the film is adequately covered, replenishment of the developer is carried out. Replenishment solutions are made up according to the manufacturers instructions and it can be the same as the solution replenished or of a less dilute nature.

The quantity of replenishment solution added should preferably not exceed 5 % of the total tank volume to keep the density variations within limit. It is also not advisable to continue the replenishment indefinitely. The developer solution should be discarded when the replenisher used is twice the original quantity of the developer.

4.2.3.4 Stop Bath

After the films are removed from the developer it is advisable to rinse them in a stop bath for about 30 to 60 seconds with moderate agitation. The purpose of stop bath or rinsing bath is

(i) to stop the action of the developer that is still adhering to the film. If the developer is allowed to act further, it would slowly start attacking the unexposed silver halide crystals. This will result in enhanced density, fog or irregular developer stains.

(ii) prevent contamination of the fixer bath by the developer solutions, thereby prolonging the life of the fixer.

A stop bath is a acid water solution. Conventional stop bath is made by mixing 100 ml of 28% acetic acid with 1 litre of water.

If a stop bath is not used, the film can be immersed in clean running water and rinsed for at least 2 to 3 minutes.

4.2.3.5 Fixing

After the film is removed from the stop bath, it is put in the fixer. The purpose of the fixer is

(i) to stop further development

(ii) to remove from the emulsion all the undeveloped silver salts thereby leaving the developed silver as a permanent image

(iii) harden the gelatin of the emulsion so that it will be more stable during subsequent processing such as washing, drying and other handling operations.

The essential ingredients of a fixing solution are

(a) clearing agent which removes the undeveloped silver halide from emulsion making it water soluble compound that can be removed by washing. The most widely used clearing agents are sodium thiosulphate or ammonium thiosulphate.

(b) a hardening agent – potassium alum

(c) a neutraliser of organic acid

(d) sodium sulphite a preservative which prevents the decomposition of the clearing agents

(e) a buffer salt which maintains the desired pH and prevents development of sludge in the fixer.

The fixer should be maintained at the same temperature as the developer and stop bath (18-24°C). The interval between placing the film in the fixer solution and the disappearance of the original diffused yellow milkiness is known as the clearing time. This is when the fixer dissolves the undeveloped silver halides. An equal amount of time is required for the dissolved silver halides to diffuse out of the emulsion and for the gelatin to harden sufficiently. Thus total fixing time should be atleast twice the clearing time. Generally a film is placed for a minimum period of 5 minutes in the fixer and agitated periodically.

> Replenishment of developer maintains constant levels of chemical activity and physical level in tank.
>
> ***Developer*** to be discarded if the amount of replenisher is twice the original quantity of the developer.
>
> ***Stop bath*** inhibits developer action and prevents fixer contamination.
>
> Fixing is the step during which
>
> * *Undeveloped silver halide is removed from the emulsion and*
>
> * *Image is made permanent*
>
> Film should be placed in fixer for a time equal to twice the clearing time.

4.2.3.6 Washing

After developing and fixing, the film must be washed well with clean running water having a temperature of about 20°C for about twenty minutes and then dried. Washing serves primarily to remove the fixing-bath chemicals. Inadequate washing results in an X-ray film with poor archival qualities. The X-ray film starts browning with age since the retained hypo will react with the silver image to form brown silver sulphide, just as silverware acquires a brown tarnish when exposed to the hydrogen sulphide produced by cooking gas. The water flow should be such that the volume of the tank is replaced four to eight times per hour.

The excess water is then removed, since the wet radiograph is not convenient to handle and easily liable to mechanical damage and the X-ray film dried.

4.2.3.7 Wetting and Drying

The second last step in manual processing is immersing the film in a wetting agent for about 1 to 2 minutes. This step is often skipped by the radiographers. When films are removed after washing, small drops of water cling to the film surfaces. If the films are dried rapidly, the areas under the drops dry more slowly than the surrounding areas. This uneven drying results in visible water marks or streaks that can interfere with the interpretation of radiographs. Immersing the films in wetting agent causes the water to drain off uniformly. This helps in preventing water marks and also ensures uniform and faster drying.

The film retains a lot of water which can be heavier than the weight of the emulsion itself. The process of drying is used to remove this unwanted water. Drying has an important bearing on the quality of the finished radiograph. It should be done without causing any damage to the emulsion or marks from uneven drying, and without exposing the damp emulsion to dust or lint. To avoid drying marks, the film can be immersed in a solution of wetting agent for about 15 – 30 seconds. This wetting agent overcomes the surface tension of water, thus preventing the formation of water droplets.

Usually the wet films are dried by blowing a current of hot air over the films or using drying cabinets. Various types of drying cabinets exist for this purpose. These cabinets often include a filter at their air intake to free the air from dust, which may stick to the wet film. Normally temperature upto 40°C is employed for drying. Some rapid drying machines are available in which films are fed and are accepted by a pair of rollers. The films come out dry in 70-90 seconds and hence offer real advantage in time. Hardening is essential in this type of rapid drier, otherwise film may stick to the rollers.

> Washing removes all fixer chemicals.
>
> Inadequate washing can cause film to brown
>
> Wetting agent helps in reducing the water spots
>
> Drying hardens the emulsion and protects it from damage during handling

4.2.4 Automatic Processors

Use of automatic film processors greatly speeds up the processing of films and minimises film artifacts. With an automatic film processing unit, the radiographer needs only to remove the exposed film from the cassette, and feed it into the processing unit. The unit then takes over and, in about 15 min, the film is ready for viewing. The advantages of course, are, that the output is increased and the film quality is improved. The automatic method is achieved by the roller transport principle (Fig. 4.2.4)

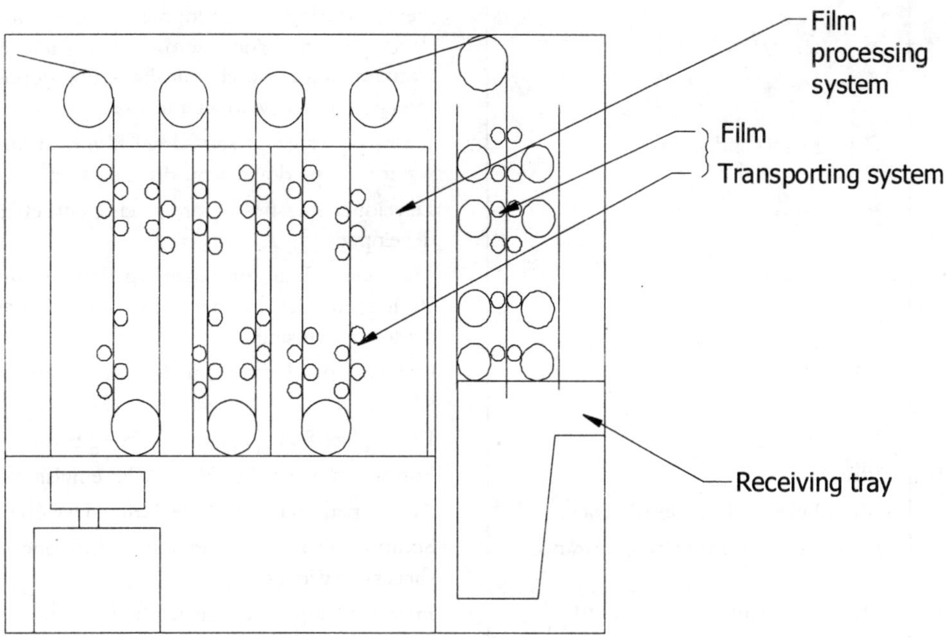

Fig. 4.2.4 Schematic sketch of automatic film processor

Table 4.2.1 lists some of the common artefacts, encountered especially during handling and processing of radiographic films. The causes and radiographic appearance are also indicated.

Table 4.2.1 Radiographic Appearance of Blemish/mark and their Causes

Sl. No.	Radiographic Appearance of Blemish/mark	Cause
1.	Black crescents	Kinking of film before processing
2.	White crescents	Sharp bending or folding of the processed film.
3.	Overall fog	Over-development, over-aged film, prolonged exposure to safe light, improper safelight filter, insufficient protection in storage from radiation sources, film stored where temperature or humidity is too high, viewing developed film, before it is properly fixed.
4.	Black streaks or Bloches	Light leaks due to faulty film holders or cassette.

Table 4.2.1 contd.

Table 4.2.1 contd.

5.	Streaks	White and black streaks at points where film was attached to hanger caused by processing solution remaining on clips from earlier use. Inadequate agitation during development. Removing film to view during development time causing developer to run across film unevenly. Contamination by chemically active deposits, interaction of developer and fixer.
6.	Drying spots and streaks	Drops of water on semi-dried film or drops of water running down semi-dried surface.
7.	Black spots	Developer splashes before placing entirefilm in developer.
8.	Brown stains	Inadequate fixing or exhausted fixing solution, Prolonged development in old developer, inadequate rinsing.
9.	Dark deposits	Oxidised products from developer orreacting silver salts
10.	Milky appearance	Incomplete fixing, exhausted fixing bath.
11.	Blisters	Formation of gas bubbles in film emulsion
12.	Air bubbles (light spots/Patches)	Air trapped on film surface during development.
13.	Reticulation (leather like appearance)	Solution too warm, extreme differences in successive processing baths.
14.	Frilling (loosening of emulsion)	Fixing in warm or exhausted bath. Prolonged washing at high temperatures
15.	Static marks	From electric discharges caused by friction between film and interleaving paper/holder
16.	White spots and areas	Pitted or worn screens. Dirt on film screens.
17.	Light fingerprints	Film touched with greasy fingers before development.
18.	Dark fingerprints	Film touched with dirty fingers before development

Questions:

Choose the most appropriate answer

1. If a film is placed in a developer solution and allowed to develop without any agitation:
   a. the radiograph will not show proper contrast
   b. it will be impossible to fix the radiograph permanently
   c. there will be a general "fogging" condition over the entire radiograph
   d. there will be a tendency for areas to affect each other.
   e. it will not have any effect on the development process
2. The selection of the proper type of film to be used for the X-ray examination of a particular part depends on:

a. the thickness of the part
b. the material of the specimen
c. voltage range of available X-ray machine
d. all of the above
e. experience of the radiographer

3. If a film is exposed to light prior to development it will most likely result in:
   a. a foggy film
   b. poor definition
   c. streaks
   d. yellow stain
   e. underexposed film
4. The four main steps in processing a radiograph are:
   a. developing, stop bath, fixing and wetting
   b. developing, fixing, washing and drying
   c. exposure, developing, fixation and drying
   d. developing, stop bath, fixing and evaluating
   e. developing, fixing, wetting and drying
5. Branched tree like or circular dark marks on a radiograph, are often caused by:
   a. film being bent when inserted in a cassette or holder
   b. foreign material or dirt imbedded in screens
   c. scratches on lead foil screens
   d. improper film handling techniques
   e. improper processing conditions
6. During manual processing of films, we sharply tap the hangers two or three timesafter the films are lowered into the developer. This is done to:
   a. disperse unexposed silver grains on the film surface
   b. dislodge any air bubbles clinging to emulsion
   c. prevent frilling
   d. agitate the developer solution
   e. do all of the above
7. The activity of the developer solution is maintained stable by:
   a. constantly agitating it
   b. maintaining processing solutions within the recommended temperature range
   c. avoiding contamination from the wash bath
   d. adding replenisher
   e. adding additional water
8. The purpose of fixation is:
   a. to remove all the undeveloped silver salts of the emulsion
   b. to leave the developed silver as a permanent image
   c. to harden the gelatine
   d. all of the above
   e. a and b above
9. For best results when manually processing film, solutions should be maintained within a temperature range of:

a. 18 to 24°C (291 to 297 K)
b. 65 to 75°C (338 to 348K)
c. 24 to 29°C (297 to 302K)
d. 75 to 85°C (348 to 353 K)
e. 4 to 24°C (277 to 297 K)

10. Water spots on films can be minimized by:
    a. the rapid drying of wet film
    b. using a wetting agent solution
    c. using a fresh fixer solution
    d. cascading water during the rinse cycle
    e. drying the film slowly using a warm jet of air.
11. The graph which expresses the relationship between the log of the exposure and the resulting density is callèd:
    a. H & D curve
    b. an exposure curve
    c. the characteristic curve
    d. a logarithmic curve
    e. a & c above
12. The normal development time for manually processing X-ray film is :
    a. 2-18 minutes in processing solutions at 24°C (75°F)
    b. 3-8 minutes in processing solutions at 16°C 60°F)
    c. 12-18 minutes in processing solutions at 68°C (154°F)
    d. 5-8 minutes in processing solutions at 20°C (68°F)
13. During manual film processing the purpose of the stop bath is to:
    a. change the exposed silver salts to black metallic silver
    b. neutralize the developer and stop the developing process
    c. eliminate most water spots and streaks
    d. none of the above
    e. help in fixing the emulsion
14. Unexposed boxes of X-ray film should be stored:
    a. flat
    b. on edge or end
    c. in a pile
    d. it does not matter where they are stored
    e. in clean rooms near radiography area
15. The density on a radiographic image refers to the:
    a. qthickness of the film
    b. thickness of the specimen
    c. weight of the film
    d. degree of film blackening
    e. the amount of silver bromide after exposure

SECTION 5

WORK PARAMETERS AND CONDITION

After reading this section you should be able to answer

- ★ *What are the parameters affecting radiographic quality*
- ★ *What is the effect of scattered radiation on radiographic image*
- ★ *What are sources for scattered radiation and their remedial measures*
- ★ *How to calculate radiographic exposures and prepare exposure charts*
- ★ *Essential requisites for viewing radiographs*
- ★ *How to select appropriate penetrameters or Image quality indicators*

Chapter 5.1

Radiographic Quality and Exposure

5.1.1 Geometric Aspects of Shadow Formation

A radiograph is a shadow picture of an object recorded on a photographic film using X-rays or gamma rays. To obtain a radiograph, therefore, we place the specimen between a source of X- or gamma radiation and the photographic film for a predetermined time. The detectability of a defect and its sharpness on the radiograph is influenced primarily by the following factors.

(i) Shape of the defect.

(ii) Orientation of the defect with respect to the direction of radiation and plane of the film.

(iii) Size of source and its distances from the defect and the film.

(iv) Position or location of the defect in the specimen.

Defects of different shapes will give rise to different types of shadows. For example a gas hole will be shown up as a circular dark patch, a crack if detected will form a dark line, a tungsten inclusion would appear to be a white patch, etc. During radiography it is always to be ensured that the object plane and the film plane are parallel to each other and the beam of radiation is incident perpendicular to the object plane. If the beam direction is not perpendicular or if the plane of the defect is not parallel to the plane of the film the shadow will be distorted.

Because of this distortion, sometimes a particular defect can produce a shadow, which may be interpreted as some other type of defect. A fine crack may be diffused altogether hardly leaving any image behind. Although before taking a radiograph one may not have any idea about the orientation of the defects present, efforts are usually made to place the film as parallel to the specimen and as normal to the direction of radiation as possible.

5.1.2 Size of the Source and its Distance from the Defect and the Film

Fig. 5.1.1 illustrates the formation of shadow of an object due to a source of finite dimension Line source. Each and every point of the line source emits radiation. The shadow cast by the defect on the film will be the resultant of overlapping shadows cast by these points, which are slightly displaced from one another. The net effect is that the shadow is diffused around the boundaries. The shadow may be divided into two portions namely

Umbra where no direct rays are incident on the film and

Penumbra – the area which is partially exposed. This gives rise to the unsharpness of the image. This unsharpness U_g is called as the geometric unsharpness and is undesirable in radiography. Mathematically,

$$U_g = \frac{f \times t}{SFD - t} \qquad \text{... (5.1.1)}$$

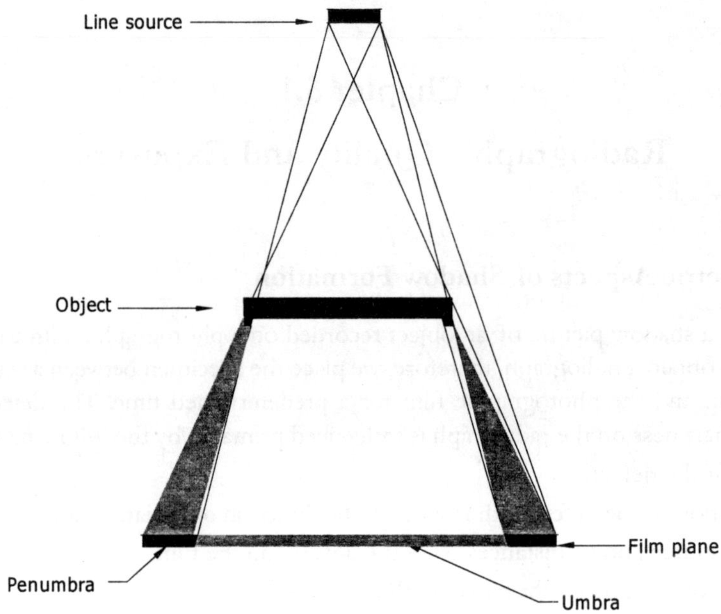

Fig. 5.1.1 Shadow of an object due to a finite source

where f is the effective focal spot or diameter of the gamma ray source, SFD is the source to film distance and t is the thickness of the object. It can be observed from the above equation that

(a) U_g can never be zero.

(b) U_g increases with focal spot size and

(c) U_g decreases with increasing source to object distance.

The magnitude of unsharpness can only be minimized by proper selection of exposure variables. Considering this fact, various codes permit a certain unsharpness value related to object thicknesses. For optimum results it is recommended that

a. source or focal spot size should be as small as practicable. Ideal source will be a point source.
b. source to object distance should be as large as practicable
c. the film should be in intimate contact with the object and
d. the source location should be such that the radiation beam is incident perpendicular to the object plane.

> Image due to line source gives rise to unsharpness. Unsharpness can only be minimised by
> * *Decreasing focal spot*
> * *Increasing source to object distance*

In a particular X-ray machine or gamma ray source, the size of the focal spot/source dimensions is fixed. Normally in all possible cases the film is placed in intimate contact with the object. Hence only the source to object distance is the factor which can be varied.

5.1.3 Image Quality in Radiography

> Radiographic sensitivity – smallest detail that can be seen on the radiograph. Depends on
>
> *Radiographic contrast and radiographic definition*
>
> Radiographic contrast refers to density difference between two adjacent areas on radiograph.

5.1.3.1 Radiographic sensitivity

The quality of a radiograph and sensitivity of flaw detection are assessed in terms of the radiographic sensitivity. Radiographic sensitivity refers to the smallest detail that can be seen on the radiograph. Quantitatively the sensitivity of flaw detection, S_f, can be defined as,

$$S_f = \frac{\text{Size of smallest detectable flaw}}{\text{Specimen thickness}} \times 100 \quad \text{... (5.1.2)}$$

This is an ideal but impractical formula, because sensitivity of flaw detection is a complex function of the size, shape, position and absorption coefficient of flaw, type of film used and density of the image obtained. It is, therefore, not possible to straightaway calculate or find the sensitivity of flaw detection. However, it is desirable to have some idea about it. Fortunately radiographic sensitivity is indicative of flaw detection sensitivity and there are means available to measure it. Radiographic sensitivity depends on radiographic contrast and radiographic definition.

5.1.3.2 Radiographic Contrast

Radiographic contrast is defined as the difference in densities between two adjacent areas on a radiograph. The greater the density difference, the greater is said to be the radiographic contrast. This is illustrated in fig. 5.1.2. Radiographic contrast is affected by two factors- subject contrast and film contrast.

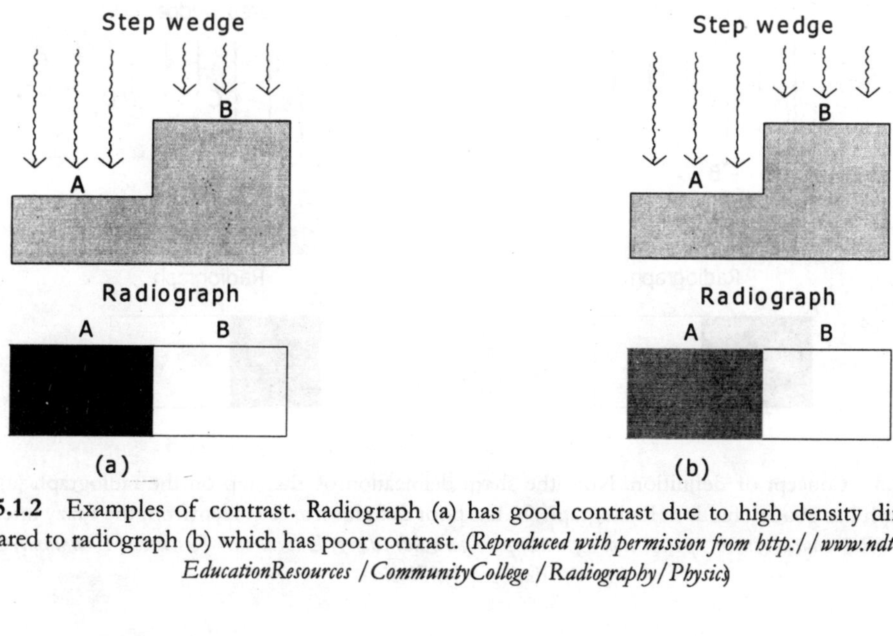

Fig. 5.1.2 Examples of contrast. Radiograph (a) has good contrast due to high density difference compared to radiograph (b) which has poor contrast. (*Reproduced with permission from http://www.ndt- ed.org/ EducationResources/CommunityCollege/Radiography/Physics*)

Subject contrast is defined as the ratio of the X- or gamma ray intensities transmitted through two different areas of a specimen. Subject contrast is affected by

Absorption differences (thickness/density) within the specimen

Quality (energy) of primary radiation

Intensity distribution and scatter radiation

A specimen with uniform thickness would have a low subject contrast while a specimen with wide thickness latitude will have a high subject contrast since the ratio of the densities would be high.

Film contrast is a measure of the response of the emulsion of the film to the intensity of impinging radiation and is a characteristic of a given film. Film contrast is affected by:

- Type of film
- Processing conditions
- Concentration of processing chemicals
- Film Density

* *Radiographic contrast depends on subject contrast and film contrast*
* *Subject contrast is affected by latitude, absorption differences within specimen, energy of radiation and scattered radiation*
* *Film contrast is affected by film type, processing and density.*
* *Film contrast is independent of the wavelength of radiation*
* *Radiographic definition refers to sharpness of image outline on the radiograph.*

Film contrast does not depend on the wavelength of radiation.

5.1.3.3. Radiographic Definition

Radiographic definition refers to the sharpness of the outline in the image. It depends on geometric factors and radiographic parameters. Fig. 5.1.3 illustrates the concept of definition.

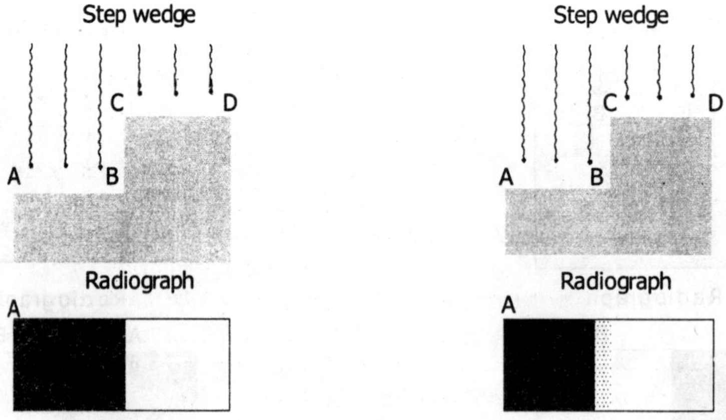

Fig. 5.1.3 Concept of definition. Note the sharp delineation of the step on the radiograph (a). This radiograph has good definition. Radiograph (b) has poor definition as is evidentfrom the fuzzy interface region (*Reproduced with permission fromhttp://www.ndt-ed.org/Education Resources/CommunityCollege/Radiography/Physics*)

The Geometric factors include

Focal spot size – For best definition this should be as small as possible.

Source to Film Distance. should be as large as possible. However, as the distance increases, intensity of radiation decreases. Hence exposure times would be large. An optimum SFD has to be considered so that exposure times are within reasonable limit.

The object should be in *intimate contact* with the film *Movement* of the object during radiography should be avoided.

> Definition depends on
>
> *Geometric parameters (focal spot, source to object distance, object to film distance) and*
>
> *Radiographic parameters (type of film, energy of radiation and development time)*

The radiographic parameters include

Type of film. Faster films have bigger grains. Hence, radiographs with faster films will have higher unsharpness.

Energy (kV) of the radiation used. As the energy of radiation is increased, the apparent graininess and hence unsharpness of the film and screen will increase.

Increased *development* of the film will increase the apparent graininess of the radiograph.

Radiographic definition sets a limit on the amount of radiographic detail that can be observed. Table 5.1.1 summarises the effect of various variables on radiographic sensitivity.

It can be seen from Table – 5.1.1 that radiographic contrast and definition are dependent on different sets of parameters and are more or less independent.

Table 5.1.1 Factors Affecting Radiographic Sensitivity

Radiographic Sensitivity/Radiographic Quality				
Radiographic Contrast		**Radiographic Definition**		
Subject Contrast	**Film Contrast**	**Geometric Unsharpness**	**Inherent Unsharpness**	**Graininess**
Radiation energy	Film type Density Development	Source to film/distance (sfd)	Radiation energy Screens	Film Type Radiation
Scattered Radiation	Fog level	Object to film distance	Screen film contact	Energy
Lead screens		Source size		
Masking Filter	Vibrations			

5.1.4 Effect of Scattered Radiation on Radiographic Sensitivity and Quality

5.1.4.1 Origin and Effect of Scattered Radiation

When a beam of radiation strikes an object, part of it is attenuated, part of it is transmitted and the balance is scattered as radiation of longer wavelength. These scattered radiations are emitted in all the directions. Some of these radiations impinge on the film thereby increasing its overall density. While the object under test itself is the main cause of scattered radiation, such radiation can also arise from the film cassette, the floor, the walls and any other object in the path of radiation. This is schematically illustrated in Fig. 5.1.4. The overall effect of scattered radiation is thus to reduce the radiographic contrast and hence decrease the resultant radiographic sensitivity.

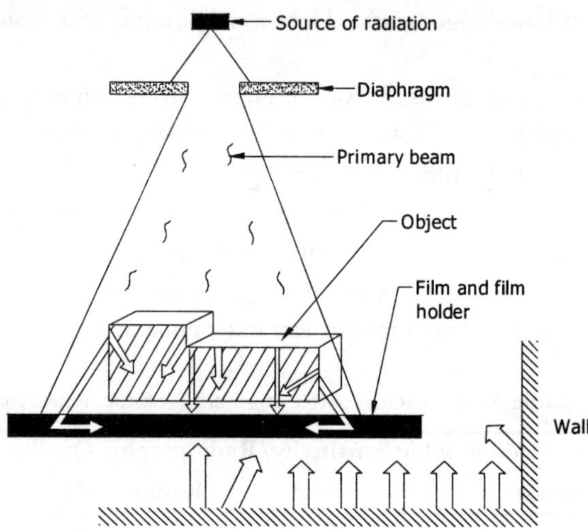

Fig 5.1.4 Sources of scattered radiation

The scattering seldom offers any problem when low kV (e.g. 40 - 100 kV) or specimens of low density material are involved. However in the mid kV range (e.g. 100 - 400 kV) and during examination of large thickness or heavy metal specimens, scattering becomes troublesome. With still more penetrating X-rays and gamma rays (energy greater than 1 MeV) the effect of scatter on the radiograph becomes less noticeable. This is due to the fact that scattered radiation is mostly in the forward direction at high energies and it, therefore, tends to assist the image formation rather than to spoil it. The amount of scattering also depends upon the shape of the specimen.

> Scattered radiation reduces radiographic contrast and hence sensitivity.
>
> Object itself is the main source of scattered radiation.
>
> Scattering predominant in mid energy range and large thickness.

5.1.4.2 Remedial Measures

Scattered radiation cannot be totally eliminated, but it can be minimised only. Given below are certain routine precautions, which are helpful to achieve the above objective.

(a) To cut down the backscatter from the floor, a sheet of lead about 3 – 5 mm thick is normally placed between the floor and the cassette.

(b) The beam should be directed and restricted with the help of diaphragms or collimators just on to the specimen or its region of interest. With this arrangement the area irradiated by the beam is restricted thereby reducing the amount of scattered radiation.

(c) The specimen and the unused portion of the film should be properly masked leaving only the areas of interest. Masking can be achieved as follows: -

  (i) For specimens of regular shape, masking can be achieved using lead sheets.

  (ii) For irregular specimens, masking can be achieved by packing a dense paste, clay lead or copper shots with a diameter of about 0.25 mm around the specimen. These shots are specially useful for radiography of irregular castings.

  (iii) Liquids, e.g. solutions of lead salts, or high density alloys of lead with low melting points may be used as masking of irregular media in certain cases.

(d) Lead intensifying screens may be used in contact with the film specially when working in mid kV range to reduce the effect of scatter. The front screen cuts the scattered radiation arising from the specimen and the back screen guards against the back scatter.

(e) The use of copper or lead filters in the path of the X-ray beam helps to reduce scatter by filtering away the comparatively softer radiation from the X-ray spectrum. This, however, reduces contrast due to hardening of the beam.

5.1.5 Evaluation of Radiographic sensitivity – Image Quality Indicator (IQI)

> Effects of scattered radiation can be minimised through the use of diaphragms, masks, filters and screens
>
> Radiographic sensitivity judged through Image Quality Indicators.

The sensitivity of a radiograph is an indirect indication of its ability to reveal flaws or thickness changes in the specimen being examined. It is, therefore, a measure of the radiographic quality achieved. The sensitivity is expressed numerically in terms of smallest detectable variation in specimen thickness as a percentage of total thickness. Radiographic sensitivity is judged through the use of image quality indicators or penetrameters.

The sensitivity or quality of radiograph is normally expressed in terms of Image Quality Indicator (IQI) or penetrameter sensitivity. IQI or penetrameters are terms which mean one and the same. These are small devices placed on the surface of the specimen during exposure. The visibility of their image on the radiograph forms the criteria of sensitivity. Detection of the images of the IQIs not only indicates that the desired sensitivity has been achieved, but it is also a pointer to the fact that appropriate parameters and technique have been adopted.

5.1.5.1 Characteristics of an IQI

The desirable characteristics of an IQI are as follows:

(i) The material of construction should be identical or radiographically equivalent to the material being radiographed.

(ii) It must be sensitive in its reading to changes in radiographic technique.

(iii) The method of reading the image of the IQI should be simple and unambiguous. Different inspectors should obtain the same value from the radiograph.

(iv) It should be versatile - i.e. applicable to a range of specimen thickness and easy to use.

(v) Since the image of the IQI has to appear on the radiograph, it should be small and it should not mask or be mistaken for flaws in the specimen.

(vi) It should incorporate some means of identification.

5.1.5.2 Types of Image Quality Indicators

A variety of penetrameter designs are being used internationally. However, the most common ones are the plaque type and the wire type penetrameters.

Plaque/Hole (ASTM) type

This penetrameter consists of a rectangular piece of metal with three holes of diameters t, 2t and 4t (Fig. 5.1.5), where t is the thickness of the penetrameter in thousandths of an inch or "thou".

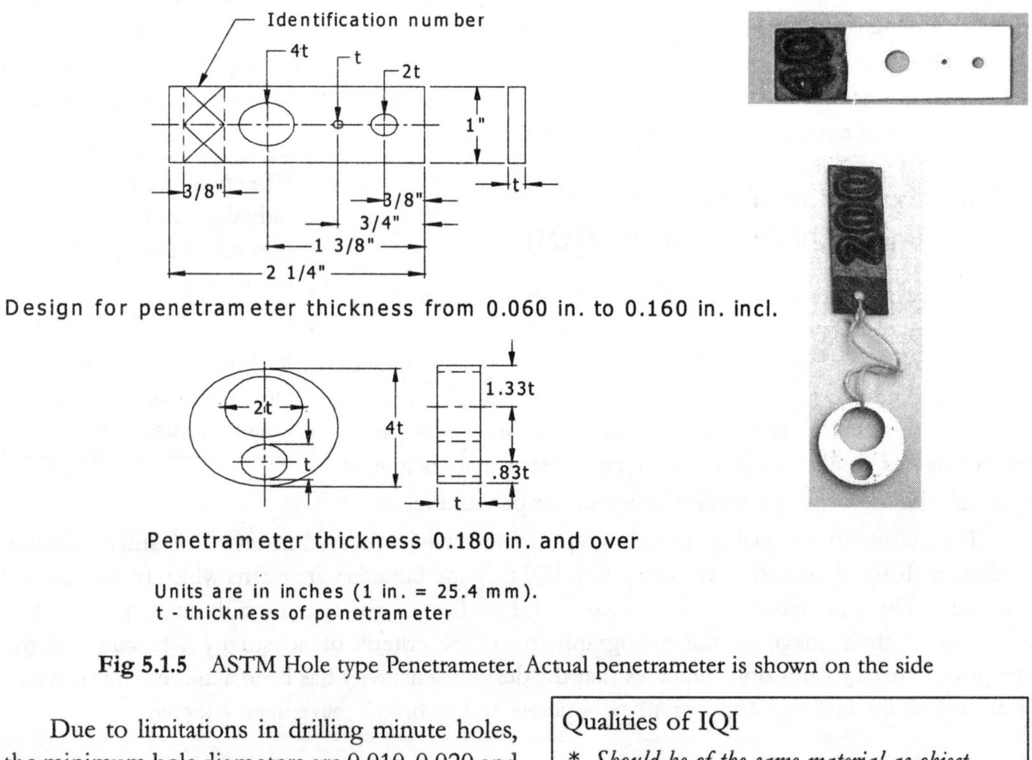

Fig 5.1.5 ASTM Hole type Penetrameter. Actual penetrameter is shown on the side

Due to limitations in drilling minute holes, the minimum hole diameters are 0.010, 0.020 and 0.040 inches respectively. As the thickness of the penetrameter increases, the diameter of the 4t hole also increases. Hence penetrameters with thickness greater than 0.180 inches are round as

> Qualities of IQI
> * *Should be of the same material as object being tested*
> * *Versatile, reliable and unambiguous*
> * *Easy to use and read*
>
> Most commonly used IQI are the plaque type and wire type

indicated in Fig. 5.1.5. Lead numbers are placed adjacent to the circular penetrameter to provide identification. Each penetrameter is identified by a lead number which indicates the thickness of the penetrameter in thousandth of an inch. The ASME Boiler and Pressure Vessel Code accept these penetrameters.

Wire type I.Q.I.

These consist of series of straight wires (at least 25mm long) of the same basic material as the specimen, mounted side by side in parallel. The distance between the axis of wires is not less than three times the wire diameter and not less than 5 mm and arranged in order of increasing diameter. The tolerance on the wire diameter is ± 5%). Wire types of IQIs have been adopted internationally. Most commonly used are DIN type and ASTM type wire IQIs (Fig. 5.1.6). The DIN type consists of 16 wires of diameters given in Table 5.1.2. The ASTM wire type penetrameter consists of sets of wires arranged in the order of increasing diameter and it covers the thickness range from 6.4 mm to 152 mm. The diameters and the corresponding wire numbers are specified in the Table 5.1.3.

Table 5.1.2 Wire Diameters for DIN Type Image Quality Indicators

Wire diameter - mm	Tolerance	Wire number
3.20	±0.03	1
2.50	±0.03	2
2.00	±0.03	3
1.60	±0.02	4
1.25	±0.02	5
1.00	±0.02	6
0.80	±0.02	7
0.63	±0.02	8
0.50	±0.01	9
0.40	±0.01	10
0.32	±0.01	11
0.25	±0.01	12
0.20	±0.01	13
0.15	± 0.005	14
0.125	± 0.005	15
0.100	± 0.005	16

Fig 5.1.6 DIN and ASTM Wire type Penetrameters

DIN IQI is standardised in three sizes. Each size is composed of 7 wires, placed parallel at 5mm distance from each other. They are 25 or 50mm long. The top is marked with DIN 62 (62 means the year of introduction of the standard) and the designation of material (e.g. Fe) and at the bottom is marked with the number of thickest wire, ISO, and number of the thinnest wire. The marking ISO means that the International Organisation for Standardardization (ISO) also adopts these image quality indicators.

Table 5.1.3 Wire IQI Sizes and Wire Identity Numbers for ASTM Type Image Quality Indicators

SET A		SET B	
Wire diameter (mm)	**Wire Identity**	**Wire diameter (mm)**	**Wire Identity**
0.08	1	0.025	6
0.1	2	0.33	7
0.15	3	0.4	8
0.16	4	0.51	9
0.2	5	0.64	10
0.25	6	0.81	11
SET C		**SET D**	
Wire diameter (mm)	**Wire Identity**	**Wire diameter (mm)**	**Wire Identity**
0.81	11	2.5	16
1.02	12	3.2	17
1.27	13	4.06	18
1.6	14	5.1	19
2.03	15	6.4	20
2.5	16	8.0	21

5.1.5.3 Placement of IQI

Plaque type IQI is generally placed parallel to the weld at a location of poorest sensitivity for a given technique. Wire IQI is placed across the weld. When the weld with the reinforcement is examined, a shim of metal should be placed under the IQI to match the total thickness.

For the placement of IQI's, the following points should be kept in mind

(i) The IQI should be placed on the source side of the specimen. If, because of specimen design or other practical difficulties, it is impossible to place the IQI on the source side of the specimen, it is probably best not to use an IQI, but if necessary, the technique to be used should be checked on a mock-up specimen of the same thickness and geometry.

(ii) The IQI must preferably be placed near the boundary of the area of interest, with the thinner step/hole IQI) or thinnest wire (wire type IQI) farthest from the beam axis.

(iii) In weld radiography, the step-with-hole type IQI's should be placed on a shim and then placed near and parallel to the weld and the wire type should be placed in

* IQI should be placed on the source side
* Plaque type IQI are placed parallel to weld near weld edge. Wire type IQI placed across the weld
* For welds with reinforcements, shims or thickness equivalent to reinforcement should be used below the IQI.
* During examination of object with varying thickness, one IQI should be used for areas in which density variation is within – 15 % to + 30%.

such a way that the wires lie across the weld length.(iv) In the case of radiography of multi-thickness castings, more than one IQI should be used corresponding to the different thicknesses in the castings

(v) For a small or complicated specimen, which does not allow the IQI to be placed on itself, the IQI can be put on a uniform block of the same material as the specimen and be radiographed along with the specimen.

(vi) One IQI should represent an area of the radiograph within which the radiographic densities do not vary by more than – 15 to + 30 percent. Otherwise two image quality indicators should be used, one showing the sensitivity level at the denser portion of the radiograph, and the second the sensitivity level of the least dense portion.

(vii) For a panoramic exposure a minimum of one image quality indicator should be used for each quadrant.

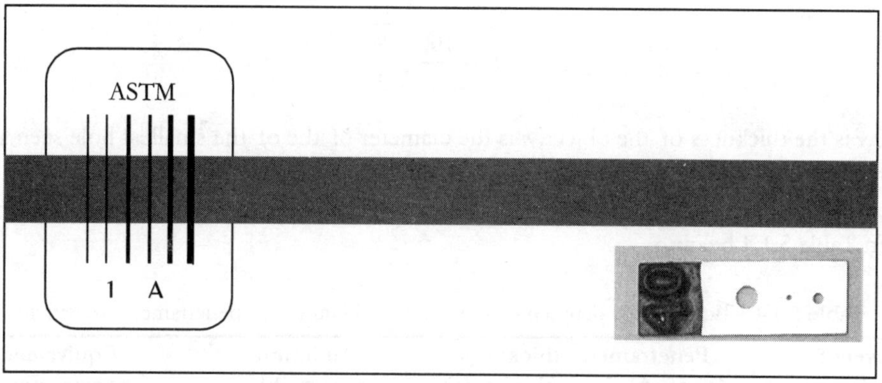

Fig. 5.1.7 Placement of ASTM and Plaque type penetrameter in a weld plate

As a general practice, the IQI must be placed on the source side of the specimen. However, when this is not practically feasible, the IQI may be placed on the film side. An IQI placed on the film side would have a higher contrast and appear to be sharper compared to an IQI placed on the source side. This is because the IQI would absorb the secondary radiations from within the object which would otherwise degrade image quality and contrast. Since the IQI is in contact with the film, the unsharpness would be lower leading to better definition. IQI placed o the film side should be identified with a letter F placed along with the IQI. Ideally the IQI should be made of the material as the object under examination. In cases where it is not possible, a material radiographically equivalent in absorption characteristics may be used.

5.1.5.4 Calculation of Radiographic Sensitivity from IQI

In case of wire type penetrameters, the radiographic sensitivity is usually assessed by the following formula

$$\textit{Sensitivity}\ (\%) = \frac{\text{Thickness}\cdot\text{of}\cdot\text{thinnest}\cdot\text{wire}\cdot\text{visible}}{\text{Specimen}\ \cdot\text{thickness}} \times 100 \qquad \text{... (5.1.3)}$$

While using eqn (5.1.3) to assess radiographic sensitivity, it is necessary to quote the type of image quality indicator. Most of the standards (DIN type, ASTM type, etc) also give their own criteria for the assessment of sensitivity.

An important point is that penetrameter is not intended for use in judging the size.

Every radiograph should have appropriate permanent identification.

The sensitivity achieved in the case of plaque type penetrameters can be expressed in two ways. One in terms of the plaque thickness and the hole that is perceived on the radiograph. For example during radiography of a 25mm thick specimen if a 0.25 mm thick plaque penetrameter is used and the 2T hole is visible, the quality level is expressed as 1 – 2T. Here the first letter represents the percentage thickness of the penetrameter in terms of the thickness of the job while the last two letters indicate the hole diameter on the penetrameter that is detected by radiography. The sensitivity of the plaque type penetrameter can also be expressed in terms of equivalent sensitivity SE as

$$SE = \frac{100}{x}\sqrt{\frac{th}{2}} \qquad \text{... (5.1.4)}$$

where, x is the thickness of the object, h is the diameter of the of the smallest hole seen and t is the thickness of the penetrameter.

Typical equivalence between the quality levels and the equivalent penetrameter sensitivity is given in Table 5.1.4 below.

Table 5.1.4 Equivalence between quality levels and equivalent penetrameter sensitivity

Level of Inspection	Penetrameter thickness (% of specimen thickness)	Minimum perceptible hole diameter	Equivalent penetrameter sensitivity percent
1-1T	1	1T	0.7
1-2T	1	2T	1.0
1-4T	1	4T	1.4
2-1T	2	1T	1.4
2-2T	2	2T	2.0
2-4T	2	4T	2.8
4-1T	4	1T	2.8
4-2T	4	2T	4
4-4T	4	4T	5.6

The quality level most normally specified for radiographic inspection is 2-2T. An important point is that penetrameter is not intended for use in judging the size or in establishing the acceptance limits of discontinuities.

5.1.5.5 Identifications

Every radiograph must have permanent identification marks which identify the job, the region examined and the orientation of the film. Lead letters and characters can be used for placement on the specimen or film to be visible on the radiograph. They should be fixed on the specimen when it is casting or weld and left in position whenever used.

However, most of the specimen will have permanent marking either by stamping, engraving or painting according to what is most appropriate for the service condition of the specimen.

5.1.6 Radiographic Exposure

The radiographic exposure is defined as the product of source strength and time for which the film is exposed to radiation. In the case of X-rays.

Exposure (milliampere-second)

= Tube current (mA) x time (sec),

and for gamma rays

Exposure (Curie-hour)

= Strength (Ci) x time (hour),

5.1.7 Methods of Exposure Determination

The determination of the correct radiographic exposure for a given specimen is very necessary to produce the required radiographic density and good quality on the radiograph. It saves labour, time and radiographic materials. The following methods may be used to determine the correct exposure.

5.1.7.1 Reference to Previous Data

Sometimes a record of previous exposures is very helpful in determining the exposure for a specimen. If a similar specimen was radiographed in the past, its exposure may be used under similar conditions. It is, therefore, recommended that a log-book be maintained by the radiographer.

5.1.7.2 Use of Characteristic curve

The characteristic curve of the film can be used to determine the correct exposures especially for samples made from mixed materials for which an exposure chart is normally not available. A trial

Practical tips to using an exposure chart

* Measure the thickness of the object under examination using a vernier or any suitable instrument.
* Select the type of film from the procedure /film selection guide provided by the manufacturer.
* Locate the relevant exposure chart corresponding to the film chosen and the type of material being tested (like Al or Steel) etc.
* In case it is an alloy or high density material for which the chart is not available, use radiographic equivalence factors to compute the thickness of the material in terms of equivalent thickness of steel.
* On the X-axis i.e. the thickness axis, locate the thickness (or equivalent thickness), of the object under examination
* Draw a line parallel to the y-axis. This line would intersect various kV lines.
* From the point of intersection draw lines parallel to X-axis. This line cuts the Y axis. The readings on the Y-axis determine the exposure milliampere – mins required to produce a density of 2.
* Since you know the tube current rating select the optimum kV that will give you a reasonable exposure time of say about 5 – 10 minutes.
* So you now know the kV, mA-mins, SFD (since chart is prepared for particular SFD), type of film, screens. You have all the parameters for taking a radiograph.

exposure (estimated from the density and thickness of the specimen material) is taken and the density thus obtained is measured. This test exposure is then corrected with the help of the characteristic curve of the film used to give the standard density of 2.0.

5.1.7.3 Exposure chart method

An exposure chart is a graph showing the relation between exposure, material thickness and energy of radiation in the case of X-rays and exposure and material thickness in case of gamma rays (for a particular isotope). For simple objects such as weld plates or castings of uniform thickness, the exposure charts provide a good guideline of the exposure parameters. However, in case of complex objects, they serve as only rough guides and appropriate corrections are to be applied. An exposure chart is normally prepared for a particular X-ray set or a particular gamma ray source. A typical exposure chart is shown in Fig. 5.1.6.

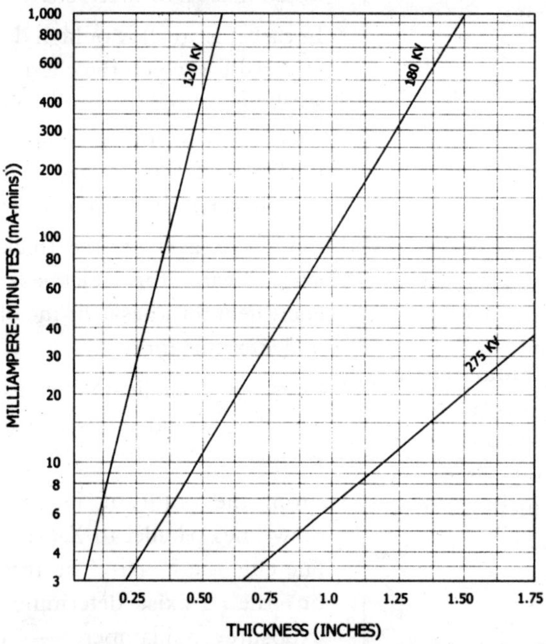

Equipment	: 420 kV(CP)
Material	: Steel
Film Type	: Brand X
Film Density	: 2.0
Screens	: Lead
Front	: 0.10 mm
Back	: 0.15 mm
SFD	: 700 mm
Processing	: Manual
(Standard Conditions)	

Fig. 5.1.8 Typical Exposure Chart for Radiography of Steel using X-rays

It can be observed that while the X-axis (thickness) is on a linear scale, the Y axis giving the exposure (mA-mins) is on a logarithmic scale. This log scale has been used primarily to compress an otherwise long scale. Any given exposure chart is valid only under a given set of conditions. These conditions are:

(a) It is valid for that particular type of X-ray machine used to prepare the chart.
(b) For a particular film type and screens (if used)
(c) For a given source to film distance

(d) Under a certain set of processing conditions

(e) For a given film density

(f) For a particular material type (eg. steel, aluminum etc.)

All these are normally indicated in the chart itself. Any change requires the application of a correction factor/remedial action. The correction factor/remedial action applying to each of the conditions listed above is discussed below

(a) *Change of X-ray Machine* – It is well known that two X-ray machines operating at the same nominal kilovoltage and milliamperage settings will give different qualities and intensities of radiation. Hence, it is advisable to prepare a new chart.

(b) *Change of Film Type* – This can be corrected for using the characteristic curves of the old and new film by determining the difference in relative exposures for obtaining the same density. In case the screens are changed from lead to salt screens, a new exposure chart may be prepared.

(c) *Change of Source to Film Distance* – The altered exposure conditions can be obtained using the exposure distance relationship.

(d) *Change of Processing Conditions* – This causes a change in the effective film speed. The appropriate correction factors are to be determined by experimentation only.

(e) *Change of film density* – If a different density is required, the correction factor can be calculated using the characteristic curve of the film

(f) *Change of material* – In case the material say steel is changed to aluminum or copper etc, with all other factors remaining same, using the radiographic equivalence factors given in Table – 5.1.5, the equivalent thickness of the material in terms of steel thickness can be computed and the new exposure factors determined.

A typical gamma ray exposure chart is given in Fig. 5.1.7. This is similar to the X-ray exposure chart and is valid for a given isotope, source to film distance, film type, film density and material.

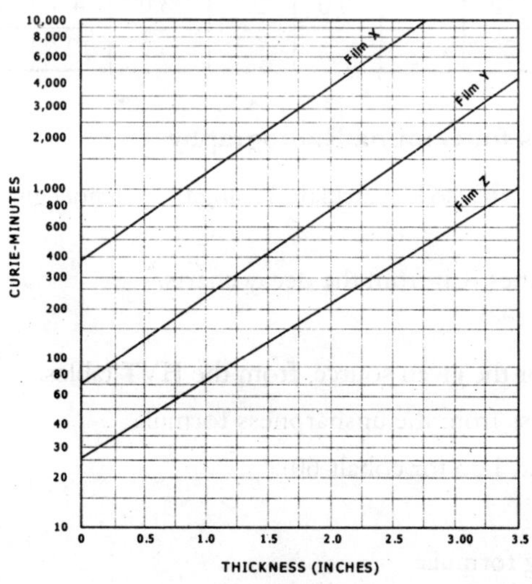

Radiation	: Ir 192
Material	: Steel
Film Density	: 2.0
Screens	: Lead
Front	: 0.10 mm
Back	: 0.15 mm
SFD	: 700 mm
Processing	: Manual
(Standard Conditions)	

Fig. 5.1.9 Typical Exposure Chart for Gamma Rays

Any change in these need corrective actions similar to those outlined earlier. A difference between this chart and the X-ray chart is that there is no variable factor corresponding to kilovoltage. Instead, there may be several parallel lines, each corresponding to film types or film density or source-film distance.

Gamma-ray exposure guides are also available in the form of linear or circular slide rules. These contain scales on which can be set the various factors of source to film distance, activity of the source etc. and the exposure time read directly.

Table 5.1.5 Approximate Radiographic Equivalence Factors

	X-rays										
Material	**50 kV**	**100 kV**	**150 kV**	**220 kV**	**400 kV**	**1000 kV**	**2000 kV**	**4 to 25 MeV**	**Ir 192**	**Cs 137**	**Co 60**
Magnesium	0.6	0.6	0.5	0.08							
Aluminum	1.0	1.0	0.12	0.18					0.35	0.35	0.3
2024 (aluminium) alloy	2.2	1.6	0.16	0.22					0.35	0.35	0.35
Titanium			0.45	0.35							
Steel		12.	1.0	1.0	1.0	1.0	1.0	1.0	1.0	1.0	1.0
18-8 (steel) alloy		12.	1.0	1.0	1.0	1.0	1.0	1.0	1.0	1.0	1.0
Copper		18.	1.6	1.4	1.4			1.3	1.1	1.1	1.1
Zinc			1.4	1.3	1.3			1.2	1.1	1.0	1.0
Brass*			1.4*	1.3*	1.3*	1.2*	1.2*	1.2*	1.1*	1.1*	1.1*
Inconel X alloy		16.	1.4	1.3	1.3	1.3	1.3	1.3	1.3	1.3	1.3
Zirconium			2.3	2.0		1.0					
Lead			14.	12.		5.0	2.5	3.0	4.0	3.2	2.3
Uranium				25.				3.9	12.6	5.6	3.4

5.1.7.4 Exposure Time Calculations for Gamma Radiography

Apart from exposure chart, exposure time can also be calculated using the formula given below in case of gamma radiography.

Calculate the present source activity in curies (Ci) from the decay charts.

Find the thickness (cm) of the object.

Find the HVT (cm) of object material, for the given source, from the HVT tables.

Find the SFD (cm) for the object thickness from the unsharpness formula.

Write RHM value (0.5 for iridium- 192 and 1.33 for cobalt-60).

Find film factor (F in Roentgen)

Substitute all these values in the following formula:

$$\text{Exposure Time (minutes)} = \frac{Fx2^{\left(\frac{X}{T_{1/2}}\right)} x(SFD)^2 x60}{CxRHMx(100)^2} \quad \text{... (5.1.5)}$$

It may be noted that this formula is not very accurate for higher thickness, particularly when using Cobalt-60 source.

5.1.8 Some Useful Formulas

The exposure charts from which the radiographic parameters are selected are prepared for a specific set of conditions such as given source to object distance. In case you wish to change any parameters, the following formulas will come in handy.

5.1.8.1 Exposure Time Relationship

Exposure = mA x time (in minutes or seconds) – for X-rays ... (5.1.6)

Exposure = Curie x time (in minutes or hours) – for gamma rays ... (5.1.7)

Using this you can determine the exposure time when the intensity is varied.

5.1.8.2 Exposure – Distance Relationship

As the distance between the source and object increases, the intensity of radiation falls as square of the distance and the exposure doubles. Mathematically,

$$\frac{E_1}{E_2} = \frac{D_1^2}{D_2^2} \quad \text{... (5.1.8)}$$

5.1.8.3 Milliamperage – Time Relationship

$$\frac{M_1}{M_2} = \frac{T_2}{T_1} \quad \text{... (5.1.9)}$$

5.1.8.4 Time – Distance Relationship

$$\frac{T_1}{T_2} = \frac{D_1^2}{D_2^2} \quad \text{... (5.1.10)}$$

5.1.9 Control of Radiographic Quality

There are many factors which contribute to the resultant quality of a radiograph. These factors should be carefully controlled by following a well recognised national/international practice usually specified in codes. Information on the following points is essential.

Technique Classification and general requirements such as very high sensitivity, high sensitivity and medium sensitivity are followed as per code.

Material and its Thickness data is essential to select the source, SFD and other important radiography parameters. Whenever possible, fabrication procedure such as welding, casting, etc. should be made available.

Film and Screen used are checked up with the recommendations of the code with respect to the technique classification.

Film Processing should be standard as per the recommendations of manufacture of the chemicals. Shortened and prolonged development time must be avoided.

Density on the Radiograph should be between 2-3. Higher density may be employed provided high intensity illuminator is available for viewing the radiographs. Fog density and expiry date of the film should be checked. Fog density should not exceed 0.2

Identification Marking System has to be explicit to enable finding the exact area of the specimen without any ambiguity. Lead markers impression must appear on the radiograph.

Source, its **active dimensions** and **calibration** data should be verified with recommendations of the code.

SFD is directly linked with specified U_g value. It also determines the beam area coverage on the specimen.

Exposure geometry with relative position of source, specimen and films shown be given in a sketch for verification.

IQI, its location bears the most important information leading to assessment of the sensitivity of a radiograph. Observed sensitivity is calculated in terms of thickness of plaque/hole or diameter of the wire discernible by eye.

Exposure Time should be in the specified range in the code.

Besides the above points, the radiograph is carefully examined on both sides with reflected light to reveal superficial marks such as scratches, crimps and other processing marks. Use of a magnifying glass would be helpful. Familiarity with film artifacts and their indication is also essential to determine the genuine discontinuity as different from the spurious indication on the film.

5.1.10 Unsatisfactory Radiographs

Table 5.1.6 lists some of the commonly encountered faults in unsatisfactory radiographs. The probable causes and the required remedial actions have also been indicated.

Table 5.1.6 Unsatisfactory Radiographs – faults, causes and corrections

Faults	Probable Cause	Remedial Action
High Density	Overexposure	View with higher intensity illuminator or reduce the exposure time suitably using characteristic curve.
	Overdevelopment	Check whether developer temperature is high
Low Density	Underexposure	Increase exposure time with the aid of characteristic curve
	Underdevelopment	Check for low developer temperature and if so increase development time. Check whether developer has become weak
High Contrast	High subject contrast due tovarying thickness of the object	Increase kilovoltage. Use multi film technique
	High film contrast	Use film with lower contrast characteristics
Low Contrast	Low Subject contrast	Reduce kilovoltage
	Low Film contrast	Use film with higher contrast characteristics
	Underdevelopment	Check developer temperature Check whether developer has depleted
Poor Definition	Specimen not in contact with the film	Keep specimen and film in intimate contact or as close as possible, if not, increase source-to film distance
	Too short SFD	Increase SFD
	Large focal spot	Use tube with smaller focal spot or increase the SFD
	Screens and film not in close contact	Ensure proper screen film contact
	Film graininess	Use finer grained (slower) film
Fog	Exposure to safelight	Check safelight lamps for correct wattage Check whether filters on the safelight are not cracked or damaged
	Improper storage conditions (located close to radiation areas or presence of excessive heat, dust, gases)	In case exposure to background radiation, store the films in lead lined boxes. In case of excessive heat etc., store the film in a cool, dry place well ventilated and free from gases and vapors.
	Developer solution	Check whether developer solution has weakened or is diluted and replace.
	Exposure during processing	Do not frequently remove the film from developer and inspect until fixing is completed

(Table 5.1.6 Contd.)

(Table 5.1.6 Contd.)

Finely mottled fog	Stale film	Replace the film
Dark patch/ streak	Defective cassette Film box opened with lights on accidentally	Replace cassette
Yellow stain	Depleted developer or Fixer	Replace the weak developer or fixer solution
	Failure to use stop Bath or to rinse	Use stop bath or rinse thoroughly
Dark branched lines and spots	Static discharge	Remove film carefully from wrapper. Avoid rubbing one film against another.
Dark fingerprints	Undeveloped film handled with chemically contaminated fingers	Use clean dry hands
Crescent-shaped light areas	Faulty film handling	Handle films carefully. Do not apply excessive pressure during handling
Light fingerprints	Undeveloped film handled with oily or greasy fingers	Use clean dry hands
Light spots or areas	Presence of dust, hair or lint between the film and screen	Keep screens clean
Reticulation (Leather grain appearance)	Temperature differences in processing solutions	Maintain all processing solutions at the same constant temperature
Frilling (Loosening of film emulsion from the film base)	Overly warm or depleted fixer solution	Maintain correct fixer temperature and replace fixer as required

Chapter 5.2
Viewing of Radiographs

The first essential pre requisite towards successful evaluation of radiographs is the availability of proper conditions for film viewing. It is of little use taking care to obtain maximum detail on the radiograph if the viewing conditions do not permit perceiving such detail. Conventional radiographs are viewed against a light source. The equipment providing the light is called as 'illuminator'. Various types of illuminator are commercially available. However, the most suitable illuminator is the one, which has a high intensity light source sufficient to penetrate films with densities up to 4, with provisions for intensity control and mechanical arrangements for masking the lighted area.

To achieve satisfactory viewing of radiographs, film density, illuminator brightness, background lighting and visual acuity of inspectors must fulfil certain conditions. These are discussed below in detail

5.2.1 Film Density

The area of radiograph under evaluation must be within the specified density range. Most of the codes and specifications require densities to be in the range of 1.5-4.0. Further the variations in the density within the radiograph must also be within a range of –15% to +30% with reference to the density measured through the penetrameter(ASTM Plaque type) body.

5.2.2 Illuminator Brightness

Recommendations for illuminator brightness are normally given for 3 film densities 1.0, 2.0, and 3.0. Values of screen brightness for intermediate values can be extrapolated. It has been experimentally observed that human eye is most sensitive to contrast difference at a brightness level of about 10 mlux. A good illuminator should be capable of producing luminance of the illuminated radiograph which is not less than 30 cd/m^2 and wherever possible approx. 100 cd/m^2 or greater. The minimum values of illuminator brightness for various densities are:

for film of density 1.0: 300 cd/m^2
for film of density 2.0: 3000 cd/m^2
for film of density 3.0: 30000 cd/m^2

The brightness of the viewing screen can be checked using a light (lux) meter and in case the same is not available, with ordinary photographic exposure meters. The exposure meter

> With a good illuminator, it should be possible to read radiographs upto a density of 4.0.
>
> Minimum value of illuminator brightness shall be 30000 cd/m^2 for a density of 3.0.
>
> Brightness of the illuminator screen can be checked using lux meters or photographic exposure meters.

should be set for a film speed of 100 ASA and held with its sensitive area close to the screen. A reading on the meter of an exposure of 0.01 of f=10 corresponds to a screen brightness of 1000 cd/m^2. Other recommendations are as follows :

a. colour of light shall be normally white, but colours between orange and pale green are acceptable

b. diffusion of light through one sheet of thin opal glass is found to be adequate.

c. Uniformity of brightness should be achieved by masking the illuminated area to the minimum required for viewing radiographic image. The edge of the radiograph should always be masked and if the radiograph contains regions of density which are considered less dense than the area under examination, these should be masked off.

5.2.3 Visual Acuity and Darkness Adapatations

The radiographic inspector should have good visual acuity i.e. the ability to see fine details and the ability to distinguish the small changes in image density. For this reason, most of the codes/ standards require that radiographers undergo eye examination atleast annually. This examination includes reading letters and numbers in eye testing charts such that

(a) distant vision should equal to 20/30 or better in at least on eye either uncorrected or corrected

(b) near vision should permit reading a minimum of jaeger no.2 or equivalent type and size letters at not less than 30 cm on a standard jaeger test chart for near vision in at least one eye uncorrected/corrected

(c) colour vision should be adequate such that the observer can distinguish and differentiate between the colours used

No matter how good an individual visual acuity is, the ability to see low contrast images depends on how well the eye is adapted to the surroundings. While no firm rules are possible, as a general guide, an observer coming from a well ventilated area should dark adapt his eyes for at least 10 minutes in very subdued lighting. The adaptation time should be a continuous period of time. Even when dark adapted, a simple method to improve the visibility of low contrast images is to lightly move the radiograph back and forth, during viewing. This is because a moving target even though it may be small is easier to pick compared to a stationary one. The visibility of faint images can also be improved by viewing the film at an angle.

Maximum distance of the eye from the radiograph should be about 40 cm. Low powered magnifying lens (3X to 4X) of reasonable area is recommended during evaluation. All film readers shall check their eye sight annually for ability to see small details at normal film viewing distance.

> Colour of illuminator light shall be normally white.
>
> The light shall be diffused with uniform intensity throughout the screen
>
> Any radiographic inspector should have good visual acuity
>
> Before interpretation, the inspector should dark adapt his eye for a minimum period of 10 minutes
>
> Maximum distance from screen to film shall be 40 cm
>
> All inspectors shall check their eyes once in a year

5.2.4 Viewing Conditions

Apart from dark adaptation viewing conditions are also very important for proper radiographic interpretation. The sensitivity of human eye is maximum when the background lighting is subdued and approximately of the same brightness as the area of interest. Hence it is necessary to restrict the area of illumination to that particular region of interest in the radiograph. Adjustable blinds or cardboard masks with holes cut out are simple methods that can be adopted for this.

Another important point is the elimination of unwanted light or glare on the radiograph during viewing. This can arise due to a variety of factors such as windows, open doors or incandescent bulbs. In general specular reflections off the surface of radiograph should be avoided. A simple method to check this is to place a white opaque card in the place of a radiograph and measure the luminance. This luminance of the card which is due to ambient light should not exceed 10% of the luminance of the illuminated film in the area under inspection. The luminance can be measured with a photographic exposure meter.

5.2.5 Interpretation of Radiographs

Successful interpretation of radiographs depends on information obtained through training or by experience and intuition. The most essential basis for sound interpretation is a wide and adequate knowledge of the manufacturing process, radiographic process and the characteristic appearance of the defects on the radiographs. While there is no substitute for experience, this knowledge can also be gained from

(a) scrutinising a wide range of specimens, radiographed and preferably correlated with sectioned samples.
(b) reference radiographs either in the form of radiographs, prints or duplicates transparencies.

> The radiographic interpretation room should be dimly lit
>
> Unwanted light, glare on the screen or directly on the eye should be avoided
>
> The key to successful interpretation need is wide field experience gained by scrutinising a number of radiographs

Questions

Choose the most appropriate answer

1. A radiograph is taken with a source to film distance of 36 inches. If the film is how placed at a distance of 46 cm from the target and all other parameters are kept the same the new exposure time would be
   a. same as in previous case
   b. longer by approximately 80 %
   c. shorter by approximately 55 %
   d. about 25 % longer than original exposure time
   e. only about 25 percent as short as the original exposure time
2. An good radiograph is obtained under given exposure conditions with a tube current of 5 mA and an exposure time of 12 minutes. If other conditions are not changed, what exposure time would be required if the X-ray tube current could be raised to 10 mA?

a. 24 minutes
b. 12 minutes
c. 6 minutes
d. 3 minutes
e. 18 minutes

3. During radiography, image quality indicators are normally placed
   a. on a shim either parallel or perpendicular to the weld region
   b. on the object facing the source of radiation
   c. on the object close to the film
   d. on a radiographically equivalent material adjacent to the object under examination
   e. all of the above
4. The primary purpose of using lead intensifying screens in industrial radiography is to:
   a. improve definition and resolution in radiographic images
   b. improve contrast in radiographic images by absorbing secondary radiation
   c. decrease exposure time
   d. b and c above
   e. all of the above
5. When radiographing to the 2-2T quality level, an ASTM penetrameter for 13mm (0.5 in.) thick 304SS alloy has a thickness of:
   a. 13 mm (0.5 in.)
   b. 5 x 10^{-5}m (2 mils)
   c. 13 x 10^{-5}m(5 mils)
   d. 25 x 10^{-5}m(10 mils)
   e. 50 x 10^{-5}m(20 mils)
6. The ability to detect a small discontinuity or flaw on the radiograph is called:
   a. radiographic contrast
   b. radiographic sensitivity
   c. radiographic density
   d. radiographic resolution
   e. radiographic definition
7. Movement, geometry, and screen contact are three factors that affect radiographic
   a. contrast
   b. unsharpness
   c. definition
   d. density e. interpretation
8. The difference in densities between two adjacent areas of a radiograph is called
   a. Radiographic contrast
   b. Subject contrast
   c. Film contrast
   d. Definition
   e. Average Gradient
9. Radiographic sensitivity, in the context of the minimum detectable flow size, depends on:
   a. graininess of the film
   b. the unsharpness of the flaw image in the film
   c. the contrast of the flaw image on the film
   d. all of the above e. none of the above
10. In order to decrease geometric unsharpness
   a. a small focal spot should be used
   b. a medium focal spot should be used
   c. The film should be as near to the source as possible

d. The object should be as near to the film as possible

e. c and d above.

11. The image of the required penetrameter and hole on the radiograph indicates that the radiograph has the required :

a. contrast
b. definition
c. sensitivity
d. latitude
e. gradient

12. A densitometer is an instrument for measuring:

a. X-ray intensity
b. film density
c. density of a material
d. tube current
e. incident intensity of light

13. Primary radiation which strikes a film holder or cassette through a thin portion of the specimen will cause scattering into the shadows of the adjacent thicker portions producing an effect called:

a. blooming
b. mottling
c. undercut
d. unsharpness
e. fog density

14. If an exposure time of 60 secs was necessary using a 1.2m SFD for a particular exposure, what time would be necessary if a 0.6 m SFD is used and all other variables remain the same?

a. 120 seconds
b. 30 seconds
c. 15 seconds
d. 2 minutes
e. 3 minutes

15. When referring to a "2t" or "4t" hole in the ASTM penetrameter, the "t" refers to the:

a. part thickness
b. penetrameter thickness
c. time of exposure
d. time for developing
e. none of the above

SECTION 6
DEFECTOLOGY

After reading this section, you should be able to answer

★ *What is meant by casting and welding*

★ *What are the possible defects that can arise during the casting or welding processes.*

★ *How these defects look like on the radiographs.*

Chapter 6.1

Defectology

6.1.1 Defect Detection in Radiography

Radiography is one of the most widely used techniques for the detection of discontinuities present in the materials. The type of material may be casting, welding and assemblies of the components. For proper interpretation of the radiographs, one should have adequate knowledge of the origin of the discontinuities and also how the discontinuities will appear in the radiographs.

The radiograph presents a 2-D image of the three dimensional discontinuity on the film. Discontinuities can be classified as linear / planar or volumetric. Radiographic examination is well suited for detecting the volumetric discontinuities. Detection of planar type of discontinuities by radiographic examination is dependent on the alignment and orientation of the discontinuity with respect to the radiation beam. Also, the detection of discontinuity by radiographic examination depends on the size and location of the discontinuity and the radiographic parameters employed. In short, detectability of defects by radiography depends upon

a. size and shape of the defect
b. its location within the object and its orientation
c. radiographic density distribution
d. radiation energy used for examination
e. processing conditions
f. viewing conditions

The types of discontinuities are classified on the basis of material manufacturing processes. Radiographic examination is mainly used on the materials manufactured by the major processes like casting, welding, forging and rolling, during heat treatment of the materials and in some cases for inservice inspection.

6.1.2 Processes

6.1.2.1 Forging and Rolling

Forging is a piece of metal shaped by hammering and or pressing the metal when it is in a hot or softened condition. The main discontinuities are forging laps and forging bursts or cracks.

Rolling is the process of rolling the metal in hot or cold condition to shape as plate or other shapes. The major discontinuities that can occur in plate rolling are the lamination, stringers and seam.

> Detectability of defects is influenced by
>
> *Size, shape. location and orientation of the defect*
>
> *Energy of radiation used*
>
> *Processing and viewing conditions*

For the detection of rolling and forging discontinuities other NDE methods such as ultrasonics are more widely used. This is because most of these defects are planar and the orientation of the discontinuities may also not be amenable for radiographic examination.

6.1.2.2 Casting

A casting may be defined as a piece of metal shaped in a mould. The mould is a patterned or shaped cavity which receives the molten metal. The process of casting involves pouring or injecting the molten metal into the mould and allowing it to solidify. The cavity or mould may be of intricate shape so that when the solidified metal is removed a part is produced which, with or without further preparation, may be used for its intended purpose. The basic steps involved in the casting process are:

a. preparation of a pattern or die to form the final shape and size of part required.
b. preparation of the mould cavity using the pattern and suitable mould materials.
c. heating and melting of the metal.
d. pouring/injecting the molten metal into the mould cavity.
e. removal and, when necessary, the further cleaning or shaping of the part.

Casting processes can be grouped into two categories (a) those using a nonpermanent mould which is used only once, and (b) those using permanent moulds in which the mould can be used repeatedly to obtain a number of casting pieces. Examples of nonpermanent mould castings include:

Sand casting

Shell mould casting and

Investment or precision casting

Permanent mould castings include:

Centrifugal castings

Die or pressure castings

Radiography is used by casting industry in two ways:

(a) As an aid in developing proper casting design. For example, pilot castings can be examined to develop proper risers, gating and chilling, and to study and control internal shrinkage.

(b) As an inspection method for determining the quality of the castings produced. Radiography can be used for inspecting casting of any of the common metals including iron, steel, aluminum, magnesium, copper alloys and zinc. Practically all sizes and shapes can be radiographed.

There are certain discontinuities and defects common to cast materials and to the different casting methods. There are other discontinuities characteristic of a particular metal or alloy or of a particular casting process. As mentioned earlier, proper interpretation of defects in the radiographic image requires knowledge of the cause of the defect. Table – 6.1.1 lists some of the common casting defects, their causes and appearance on the radiographs. Fig. 6.1.1 is the digitized radiographic images of some of these casting defects.

Table 6.1.1 Typical Casting Defects – Causes and Radiographic Appearance

Defect	Cause	Radiographic Appearance
Cold shut	Failure of molten metal to unite/fuse with solid metal.	Well defined continuous or intermittent dark lines.
Hot tears	Formed due to stresses setup during solidification of the metal. Most likely to occur near areas joining thick and thin sections.	Filamentary or ragged dark lines interconnected or discrete and of variable width.
Shrinkage	Lack of enough molten metal to fill when metal contracts.	Dark areas with irregular outline.
Gas and Blow hole	Discrete cavities of 1.6 mm dia. or more formed due to entrapment of gas in the molten metal.	Dark (mostly) rounded indication with smooth outlines.
Inclusions	Entrapment of materials such as dross and sand.	Dark or light areas of irregular shape with sharp edges and folded appearance.
Cracks	Formed due to stresses setup during solidification of metal.	Dark lines straight or curved continuous or intermittent occurring singly or branched.
Core Shift	To produce internal cavities in the casting cores (usually sand) are used. If the cores are not fixed firmly, when the molten metal poured this gets shifted.	On the radiograph the cored-out section is darker than the normal metal thickness, can be easily metal can be easily identified.
Misrun	Misrun occurs when the molten metal fails to fill the mold cavity completely.	Irregular dark areas with smooth outline.
Un-fused Chaplets	Chaplets are metallic supports/spacers used to maintain the core in position. Usually of the same metal as that of castings.	Appears as a dark outline with a shape similar to that of the chaplets.
Segregation	Improper diffusion of the alloying elements used for casting.	This appears as dark or light (depending on their density), streaky area of irregular shape.
Mottling	Caused due to X-ray diffraction due to large grain size.	Radiographs with adjacent Dark and light spots or lines.

6.1.3 Welding

This is one of the oldest and most widely used materials joining process. The American Welding Society definition for a welding process is "a materials joining process which produces coalescence of materials by heating them to suitable temperatures with or without the application of pressure or by the application of pressure alone and with or without the use of filler material". Though the science and art of welding is more than 100 years old, weld continues to be the weak link in a structure. Radiography is an indispensable tool in the examination of welds. Interpretation of weld radiographs requires a certain basic knowledge of materials joined, type of

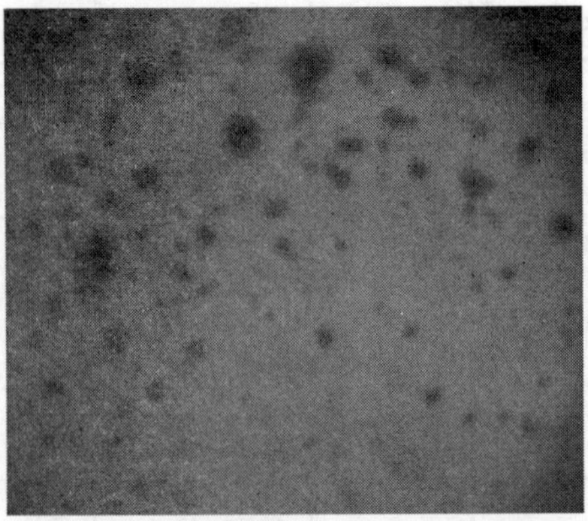

(a) Gas voids

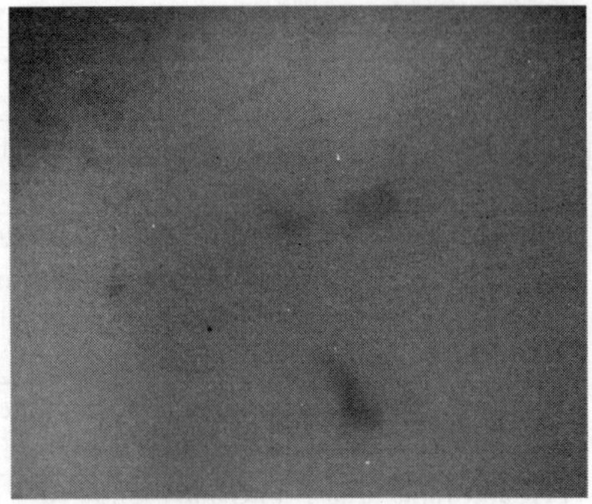

(b) Inclusions

(c) Hot tear

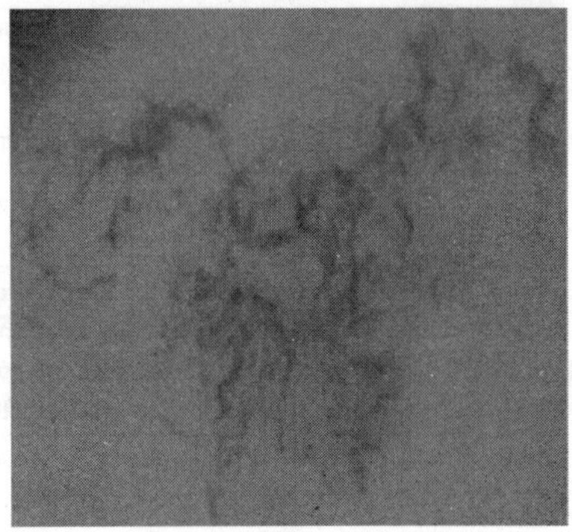

(d) Shrinkage

(e) Crack

(f) Unfused chaplet

Fig. 6.1.1 Digitised radiographic images of some typical casting defects

joint, welding processes, probable discontinuities that can arise and the radiographic appearance of these discontinuities. These points are just covered briefly in the following sections.

6.1.3.1 Materials

Generally same materials are joined together by welding. In some special cases dis-similar welding is involved to get required properties (Ex: Carbon steel to alloy steel/stainless steel). Also, in other cases certain materials are overlayed on a base material (Ex: Stainless steel overlay on carbon steel, stellite overlay on stainless steel).

6.1.3.2 Welding processes

Welding can be classified as either "manual", "semi-automated" and "automated". These terms refer to the method of control and guidance. In manual welding, the entire welding operation is performed and controlled by hand. In semi-automated welding, the welding operation is manually controlled but the equipment controls the filler metal feed. In automated welding, the weld is made with equipment that controls and performs the entire welding operation without constant observation and adjustment of the controls by an operator. There are various types of welding processes such as Shielded Metal Arc Welding (SMAW), Gas Tungsten Arc Welding (GTAW), Gas Metal Arc Welding (GMAW), Submerged Arc Welding (SAW) and Electro-Slag Welding. Some of the newer processes include Electron Beam Welding (EB Welding), laser welding, friction welding and plasma arc welding.

6.1.3.3 Joint Types

There are five basic types of joints. These are the butt joint, edge joint, fillet or tee joint, corner joint and the lap joint. A number of variations of these are possible depending on the manner of preparation and assembly. While butt joints can be readily radiographed, corner and fillet joints pose difficulties.

Inspection of welds is one of the major applications of radiography. In addition to having some knowledge about the intended purpose of a product, for correct interpretation and evaluation of the defects a radiographer is also expected to have a basic knowledge of the material, welding type, welding terms and defects peculiar to welding and how they are caused. Typical discontinuities encountered during these welding processes, their causes and radiographic appearance is summarized in Table – 6.1.2. Fig. 6.1.2 is the digitized radiographic images of welds with some of these welding defects.

Table 6.1.2 Welding Defects-Causes and Radiographic Appearance

Defect	Cause	Radiographic Appearance
Arc strike	Due to localized heat	Star shaped cracks
Burn through (Icicles)	Results when too much heat causes excessive weld metal to penetrate the weld zone. Lumps of metal sag through the weld creating a thick globular condition on the back of the weld.	Appears as dark spots surrounded by light globular areas.

Table 6.1.2 contd.

Table 6.1.2 contd.

Central line crevices	Valleys in between two passes	Continuous dark likes.
Cracks (cold)	Rupture of metal under stress below 400°F.	Dark well defined lines
Cracks (hot)	Rupture of metal under stress above 400° F., usually associated with grain boundary.	Dark lines, some tree like branching.
Cracks (base metal)	Hardening of base metal in weld area, base metal.	Dark defined lines from weld reinforcement into base metal
Incomplete fusion	Improper fusion between weld and parent metal, usually with trapped weld	Dark lines usually running parallel to weld.
Lack of penetration	Failure of penetration due to poor gap,. low current	Dark continuous line in the middle of the weld
Gas inclusions	Entrapment of gas due to wet electrode or inadequate regulation of current.	Dark round or elongated images with defined outline.
Concavity	Normally in the root due to improper movement of electrode or too much heat in II pass over I pass.	Localized dark areas in the weld region.
Weld mismatch/ offset	Associated with a condition where two pieces being welded together are not properly aligned.	A noticeable difference in density through the weld length between the two pieces.
Slag inclusion	Entrapment of nonmetallic solid materials.	Dark irregular shapes may be continuous with sharp edges.
Tungsten inclusion	Entrapment of tungsten from tip of electrode in TIG process.	White area distinctively seen in the weld.
Undercut	Use of high current, which burns edges on parent metal.	Dark continuous line, following the edge of the weld.

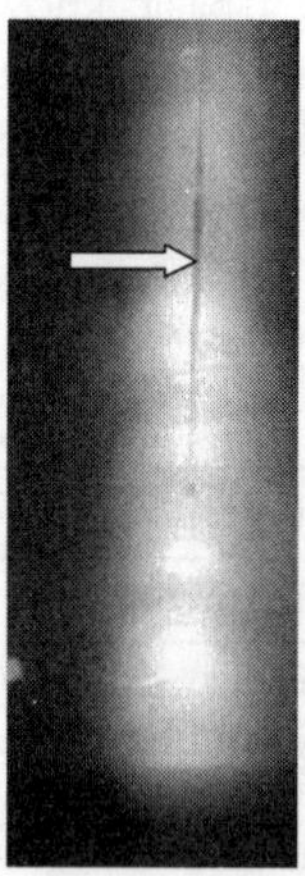

(a) Lack of penetration

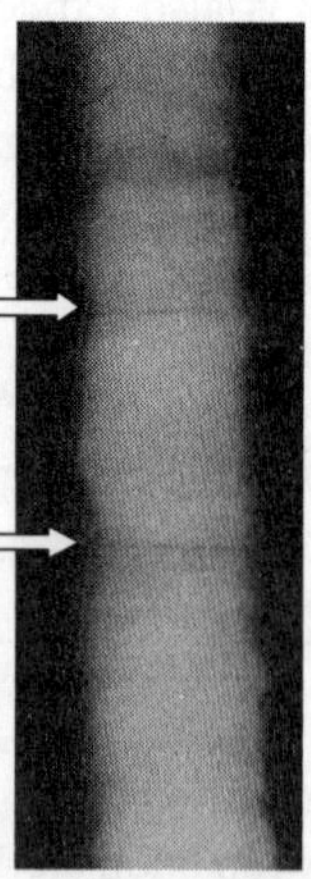

(b) Transverse crack

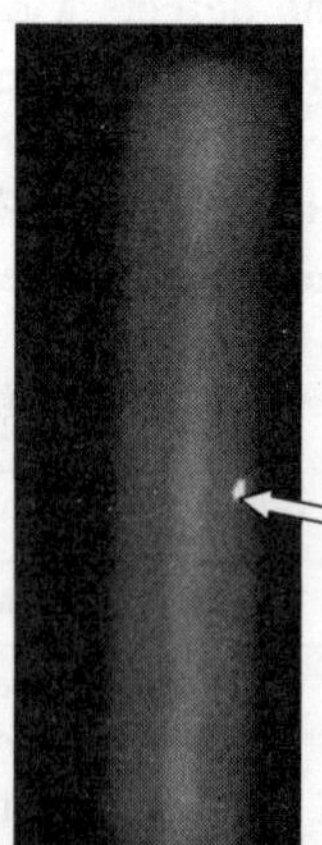

(c) Tungsten inclusion

(Fig. 6.1.2 Contd.)

(Fig. 6.1.2 Contd.)

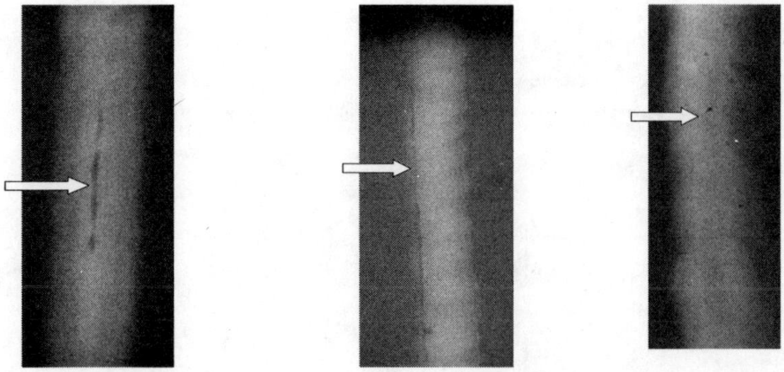

(d) Slag line **(e) Lack of side wall fusion** **(f) Scattered porosity**

Fig. 6.1.2 Digitised radiographic images of some of the commonly occurring weld defects

SECTION 7

SELECTION OF RADIOGRAPHY TECHNIQUES

After reading this section, you should be able to answer

- ★ *Identify the right parameters for radiography*
- ★ *Select appropriate technique*
- ★ *Select the best shooting angle for radiation beam*
- ★ *Position the film appropriately for best sensitivity*

Chapter 7.1

Selection of Radiography Techniques

7.1.1 Introduction

A good radiograph will have minimum distortion, sharp definition, high contrast, adequate density when geometric and exposure parameters are controlled. These four requirements influence all of the techniques of radiography. Any technique that fulfills one or more of them, without compromising on any other, is a good one. Correct film processing (described in Section-4) is an essential part of good radiographic practice. An ideal exposure technique can result in a worthless radiograph if the film has not been handled properly during processing. Thus, a radiographer needs to follow the procedures that meet the requirements of a good technique and also adopt proper film processing procedures to produce a good quality radiograph.

The ability to develop correct radiographic procedures or techniques for parts of any size is primarily gained through experience. There are however, certain fundamental points to be considered. These include:

(a) The correct film, screen and radiation energy must be used.

(b) Scattered radiation must be understood and measures to be taken identified for controlling it.

(c) Factors that affect contrast and definition must be understood and procedures should be employed to obtain optimum radiographic sensitivity.

The variables that affect radiographic sensitivity have been discussed in Section-5. Before the proper radiographic technique or procedure can be determined, the following factors need to be considered:

(i) Information on the nature of the test object - material, thickness and configuration, end use of the part, service requirements and conditions, areas of stress etc.,

(ii) Fabrication process and anticipated location and nature of discontinuity

(iii) Any specification requirement that must be followed.

(iv) Area of inspection, coverage and minimum number of exposures needed to provide a complete assessment of the object

(v) Determination of equipment, accessories, orientation and exposure conditions.

Factors (i) and (ii) are general requirements and related to materials and basic metallurgy. This chapter presents information on proven shop and laboratory exposure techniques. Radiography techniques can be affected by the following parameters: (a) radiation source (b) Source to film distance (SFD) (c) films used and (d) Physical arrangement.

7.1.2 Choice of Radiation Source

Once radiography has been selected as the test method, the choice of radiation source narrows down to X- or gamma radiography. Ideally, there is a best equipment selection for any radiographic test. Practically however, most radiography is accomplished by using the equipment available. Because of its flexibility, ease of operation, ability to vary the energy, higher intensity, better contrast radiographs and lesser radiation hazards, X-radiography is preferred over gamma radiography. Gamma radiography is usually selected for industrial applications that involve:

a. high radiation energy requirements.
b. low testing rates.
c. simultaneously exposure of many specimens.
d. confined areas where x-ray cannot be used.
e. field inspections in areas where electrical power is difficult to obtain and
f. tasks where time is not a consideration.

7.1.2.1 Source Size

Source size is a factor in every radiograph and is of primary consideration when purchasing a X- or gamma ray equipment. X-ray focal spots range from 0.1 mm to a 5 mm while in case of gamma ray sources the source size is of the order of few mm. Depending on the overall requirements, the X-ray equipment or gamma ray source is selected. Once equipment is available, the source size is fixed and the radiographer has no option but to consider other factors such as the source to film distance.

7.1.2.2 Source to Film Distance (SFD)

In selecting a source to film distance, apart from the source size (which is a constant depending on the type of equipment) three other factors need to be considered namely the penumbral effect (geometric unsharpness-) object thickness and object to film distance. These are related as

$$U_g = \frac{ft}{SFD - t} \qquad \text{... (7.1.1)}$$

where f is the source size, t is the distance between the top surface of the object to the film and SFD is the source to film distance. When the film is placed in intimate contact with the object, t is taken as the thickness of the object. Codes normally specify the maximum permissible geometric unsharpness. Hence based on this and the thickness of the object, the minimum source to film distance can be calculated. It should be highlighted that increasing the SFD for a given object thickness reduces the geometric unsharpness and improves radiographic definition. Hence a high SFD is preferred. However, since increasing the SFD decreases the intensity of radiation reaching the film (inverse square law), exposures get longer. Hence a compromise needs to be worked out for an optimum SFD which will ensure

> X-radiography is preferred over Gamma radiography due to
>
> - *Ability to vary X-ray energy*
> - *higher intensity*
> - *better contrast radiographs*
>
> Geometric Unsharpness is reduced by
>
> - increasing SFD
> - decreasing focal spot

reasonable exposure time. SFD's vary from 400 mm to about 1000 mm. Generally, as a rule of thumb, radiographers use an SFD that is not less than eight to ten times the object thickness. SFD also controls the area of the specimen covered in a single exposure.

7.1.2.3 Type of Film and Screens

Film characteristics have been dealt with in Section – 4. With most industrial radiographic films, the same degree of contrast can be obtained regardless of the film speed. This is because the characteristic curves of industrial X-ray films are similar in shape. Thus, the degree of sharpness required in the radiograph fixes the speed of the film that is acceptable. Choice of film ranges from class I to IV. Generally for detection of finer defects, fine gain high contrast films are used. Most of the film manufacturers provide a guide to the type of film to be used for a given thickness range. Lead screens are most widely used due to their ability to

(a) absorb longer wavelength scattered radiation and (b) intensify the primary beam thus reducing exposure times. Flurometallic and fluorescent screens are seldom used due to their higher screen unsharpness. Normally the front screens are thinner compared to back screens.

7.1.2.4 Physical Arrangement

The physical setup of the source, object and film during exposure affects the radiographic quality. The radiation beam from the source should be aligned normal to the region of interest. The film should be as close to the object as possible. Only the region of interest should be exposed using appropriate collimators or masks or diaphragms. This helps in reducing the scatter radiation. Scatter radiation reduces the overall radiographic contrast and hence the sensitivity. A filter between the source and the object or between the object and the film also helps in reducing scattered radiation.

7.1.2.5 Exposure Conditions

> Lead screens are widely used
>
> Fluorescent screens are not used due to high unsharpness
>
> Exposure factors for X-radiography is kV, mA and time and Gamma radiography is source strength, energy and time

In the case of X-radiography, kilovoltage, milliamperes and time are the exposure factors. The choice of kV is based the thickness of the specimen. This can be determined using the energy. vs. thickness graphs provided by codes such as ASME or by experience. Using the exposure charts, one can determine the exposure values. Exposure charts as mentioned in Section – 5 are for a particular SFD, film and material. For changes in SFD, the inverse square law can be used to compute the new exposure times. Exposure charts combined with the film characteristic curves are used for getting the correct exposure values in case of change of film. In case of change of material (for example steel to copper) the exposure factors can be obtained by combining the exposure chart information and the radiographic equivalence factors.

In case of gamma radiography, source energy, source strength and time are the exposure factors. Source energy is a function of the radioisotope and is a constant. Isotopic sources decay with time. Hence, source strength must be known at the time of exposure. Since source strength and time are reciprocal functions, the length of exposure time is determined by the source strength. For changes in SFD, film or material, the information contained in the gamma ray exposure chart coupled with the inverse square law, film characteristic curves or radiographic equivalence factors can be used to compute the new exposure factors.

7.1.3 Weld Inspection Techniques

The exposure arrangements discussed in the following paragraphs and illustrated in the figures are commonly used. These arrangements are common for either X-ray or gamma ray sources and the basic principles of film density and contrast as related to radiographic parameters such as source energy, SFD and exposure apply to each of these arrangements.

Fusion welds can be on flat plates or cylindrical objects, such as pipes. The weld can also be of butt or fillet type. Radiographic inspection of these welds with sensitive techniques reveals flaws, such as cracks, lack of root penetration, lack of fusion, porosity, inclusions, etc. Occurrence of such defects in weld can generally be identified with joint preparations, welding procedure, heat treatment, etc. Information on all these basic factors guides in anticipating the occurrence of flaw at a location and thus, helps in selecting particular exposure geometry.

In all cases of radiography, to ensure the correct degree of sensitivity, the radiographer should select the most suitable penetrameters. Placement of the penetrameters should be done appropriately so that the image of the penetrameter is clearly visible on the radiograph. In the case of welds with reinforcements or backing strips, appropriate shims should be used to compensate for the excess reinforcement.

7.1.3.1 Welded Flat Plates

Figures 7.1.1 and 7.1.2 illustrate flat weld areas. This type of weld can be easily radiographed due to its simple and straightforward geometry. Being of more or less uniform thickness, the subject contrast is small and exposure calculations are easy. Fig. 7.1.1 is a common butt weld. An exposure angle of 90° would be adequate for this joint. Fig. 7.1.2 is an example of a single V groove butt weld. In this case, apart from the 90° exposure angle, an angular exposure is also suggested for detection of lack of fusion between the weld and parent metal.

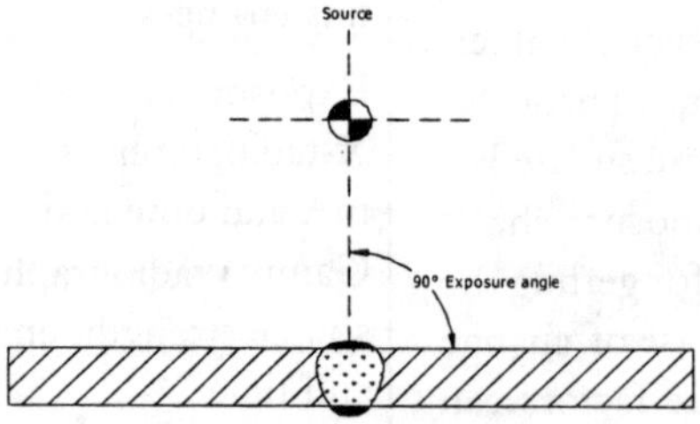

Fig. 7.1.1 Common Buttweld

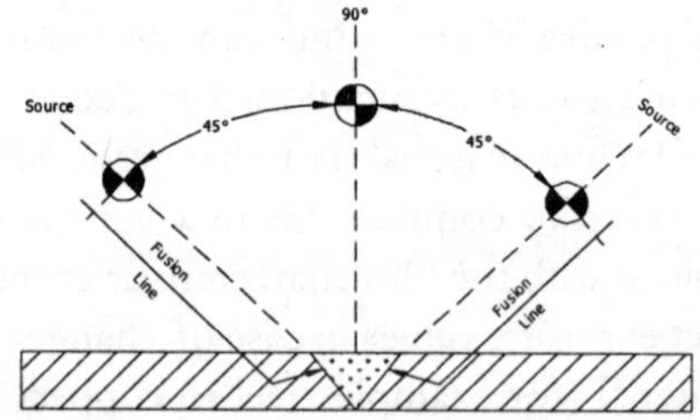

Fig. 7.1.2 Single V groove Buttweld

7.1.3.2 Fillet Welds

Fillet Welds are generally made with square or bevel edge preparation. The exposure setup for joints with 100% corner penetration and 100% standing leg penetration are shown in Figs. 7.1.3

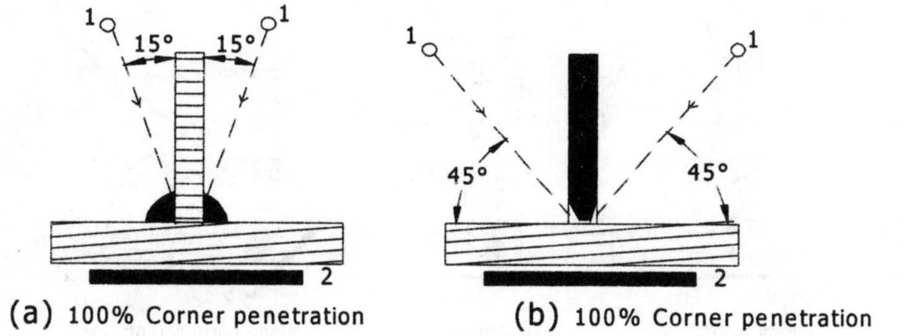

1. Source *2. Film*

(a) 100% corner penetration *(b) 100% standing leg penetration*

Fig. 7.1.3 Radiography techniques for fillet welds

(a) & 7.1.3 (b) respectively. The source is positioned at an angle varying between 15° – 20° in case of 7.3 (a) and about 45° in case of fig. 7.1.3 (b) to enable detection of lack of side fusion and also lack of root fusion.

For the examination of square joints with four fillets and lap joints the radiographic technique is illustrated in Fig. 7.1.4 (a & b).

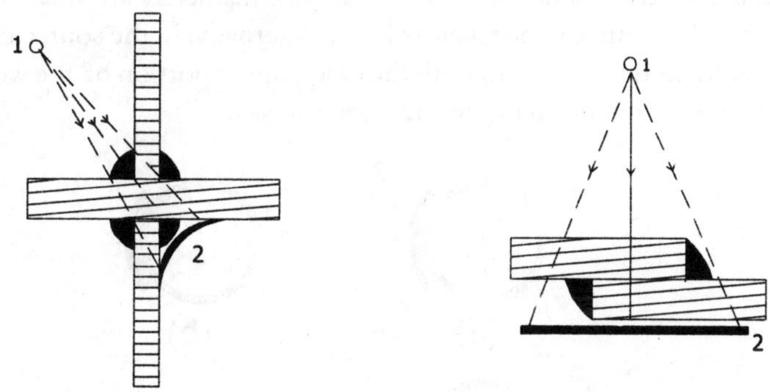

(a) Square joint **(b) Lap joint**

1. Source *2. Film*

Fig. 7.1.4 Radiography techniques for square and lap joints

Corner joints may have configuration as shown in Fig. 7.1.4. The radiographic technique for their examination is illustrated in Figs. 7.1.5 (a-c).

7.1.4 Technique for Welded joints on Pipes or Cylindrical Objects

Depending upon the size and accessibility of both sides of the pipe, the following techniques can be adopted in the case of pipe welds.

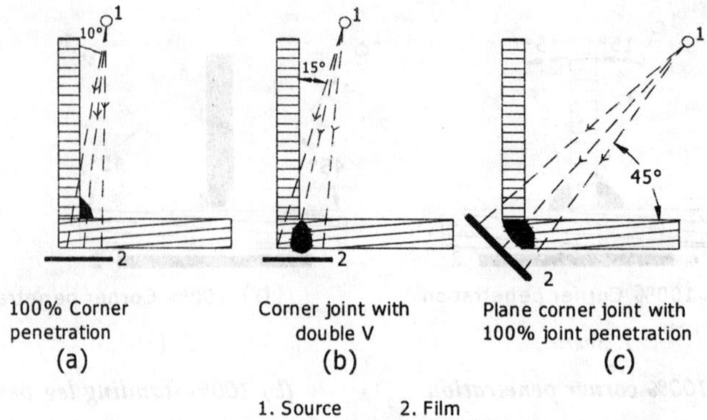

Fig. 7.1.5 Radiography techniques for corner joints

7.1.4.1 Single Wall Single Image Technique

In this technique, several methods of exposure are possible (Fig 7.1.6). The ideal position to locate the source would be the center of the pipe or cylinder as in (a) & (d). The films can then be wrapped around and in a single panoramic exposure the weld can be covered.

However, if the center is not accessible or if pipe diameters are small, the source can be off the center and the exposure can be taken as in (b). Alternatively, the source can be located outside and the film placed inside as in (c). In both the cases, only a portion of the weld would be covered. The entire weld is covered by taking several exposures.

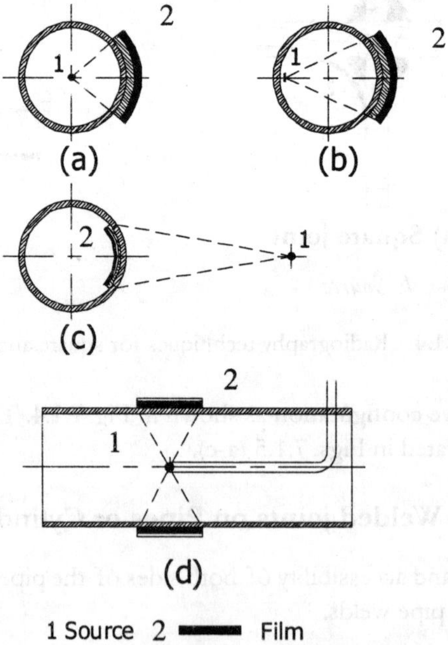

Fig. 7.1.6 Techniques for pipe exposures

7.1.4.2 Double Wall Techniques

If there is no access to the inner side of the pipe, to keep either the film or the source, double wall techniques may be adopted. Double wall techniques can be classified as double wall single image and double wall double image techniques.

Double Wall Double Image (DWDI) technique is adopted for pipes with OD less than 89 mm. This is illustrated in Fig. 7.1.7 (a). In DWDI, the source is offset by angle with respect to the weld, so as to avoid overlap of the top and bottom weld. Generally, 10° - 15° offset angle is given for the exposures. Two exposures are taken by rotating the pipe through 90°.

Double Wall Single Image technique is adopted for pipe welds with OD greater than 89 mm. This is shown in fig. 7.1.7 (b). The source can be positioned on the pipe or away from the pipe, depending on the diameter of the pipe, SFD required and accessibility.

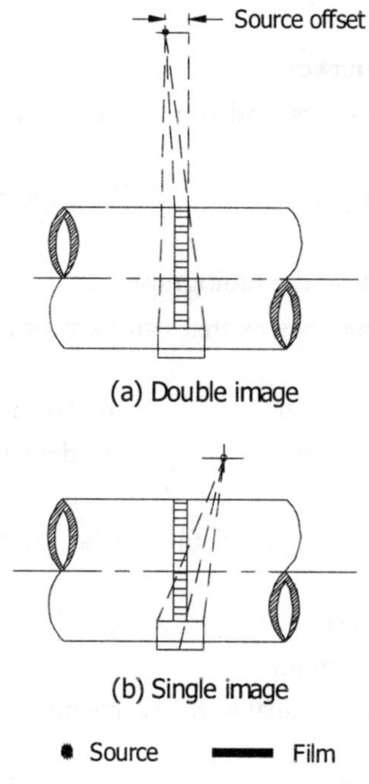

Fig. 7.1.7 (a) Double Wall Double Image and (b) Double Wall Single Image Technique

7.1.5 Radiography of Complex Parts

When complex parts are to be radiographed, it is often advisable to zone the areas and individually radiograph them. Careful study of the positioning of the film and beam of radiation can reduce the number of exposures, with resultant economy and more important, can provide the difference between adequate and inadequate sensitivity.

When solid cylindrical sections such as a solid round bar or casting section is to be examined one or two radiographic exposures may be necessary. If with a single exposure, edge loss due to undercutting occurs, it may be necessary to take two exposures with the bar turned an angle of

90° between exposures. The loss of the edge of the casting image due to undercutting or magnification factors can be reduced by forming the film to the shape of the part.

7.1.6 Double Film technique

There is a limitation on the specimen thickness range that can be inspected satisfactorily in a single radiograph. One method of extending this thickness range and thereby reducing the number of exposures required for a particular specimen involves the simultaneous exposure of two films of different speeds – one slower and other faster. With proper selection of films and exposure conditions, thicker sections will be recorded on the faster film, while the thinner sections would be recorded on the slower film. The double film technique can be used with or without lead screens.

Questions

Choose the most appropriate answer

1. Source size, specimen thickness, and source-to- specimen distance are the three factors that determine the:
   a. density of the radiograph.
   b. exposure of the radiograph
   c. film size
   d. unsharpness recorded on the radiograph
2. The range of specimen thicknesses that can be properly recorded on the radiograph is referred to as :
   a. sensitivity of the radiograph
   b. latitude of the radiograph
   c. accuracy of the radiograph
   d. density range of the radiograph
   e. all of the above
3. The selection of the proper type of film to be used for the X-ray examination of a particular part depends on:
   a. the thickness of the part
   b. the material of the specimen
   c. the voltage range of the available X-ray machine
   d. all of the above
   e. a & c above
4. In order to decrease geometric unsharpness:
   a. radiation should proceed from as a small area on the anode
   b. radiation should proceed from a large area on the anode
   c. the film should be as far as possible from the object being radiographed
   d. the distance from the anode to the material examined should be as small as practical
   e. c and d above
5. When radiographing a weld it may be advisable to grind the surface to a smooth or at least clean surface. The purpose of this is to
   a. reduce exposure time
   b. ensure uniform density over the total surface

c. reduce chance of missing defects
d. improve film contrast
e. make the weld appear acceptable

6. The focal spot size of an X-ray machine must be known in order to determine the:
   a. geometric unsharpness value
   b. kilovolt peak output
   c. milliamperage settings
   d. exposure time
   e. select the appropriate source

7. Which of the following factors will not materially influence the image density of a radiograph?
   a. the type of film used
   b. the size of the film
   c. the total amount of radiation emitted by the X-ray or gamma ray source
   d. the intensifying action of the screen

8. Lead screens are put in direct contact with the film to:
   a. increase the photographic action on the film
   b. absorb the longer wavelength scattered radiation
   c. intensify the photographic effect of the primary more than the scattered radiation
   d. do all of the above

9. The correct equation for determining geometric unsharpness Ug is:
   a. $Ug = -f/td$
   b. $f = U_g = t/d$
   c. $U_g = dt/f$
   d. $Ug = ft/d$

   where d = source-to-object distance; t = object-to-film distance
   f = size of radiation source

10. The technique to be adopted for radiography of a weld in a pipe of OD 79 mm and wall thickness 6 mm is
    a. Single wall single image technique
    b. Double wall double image technique
    c. Double wall single image technique
    d. Single wall double image technique
    e. a and b above
    f. Any of the above depending on source location

SECTION 8

CODES AND STANDARDS

After reading this section, you should know

- ★ *What is meant by Code, Standard or Specification.*
- ★ *Commonly used terminologies in standardisation*
- ★ *What are the important codes and standards (pertaining to radiography) that are widely used internationally.*

Chapter 8.1
Codes and Standards

8.1.1 Genesis of Codes and Standards

In the last century, the engineering structures and equipment were made based on imagination and experience of an architect or manufacturer. The life of the structure or equipment depended on the intelligence and skill of the constructor. There were no set rules or guidelines for safe design and fabrication practices. Due to this a number of boilers failed in a catastrophic manner leading to loss of lives and property in America. Till 1905 people used to think that Boiler explosions are either an inevitable evil or "an act of God".

On March 20, 1905, a disastrous boiler explosion occurred in a shoe factory in Brockton, under Massachusetts State of America. The explosion killed 58 persons, injured 117 persons and damaged property worth a quarter million dollars. This catastrophic accident made the people of Massachusetts to realize the necessity and desirability of legislating rules and regulations for the construction of steam boilers in order to secure their maximum safety. After much debate and discussion, the state enacted the first legal code of rules for the construction of steam boilers in 1907. It was subsequently realized by others as well that accidents could be prevented by the proper design, construction and inspection of boilers and pressure vessels. In 1911 the Council of the American Society of Mechanical Engineers ASME) took up the task of drafting the first rules for construction of stationary boilers and for allowable working pressures. Known as the ASME rules - 1914 edition, this was adopted in the spring of 1915. Today ASME is one of the most widely used boiler and pressure vessel code. Since 1914 many changes have been made and new sections added to the code as the need arose.

8.1.2 Definitions of Some Common Terminologies

(i) Quality

Quality of an industrial product is defined as the "fitness of the product" to do the job required by it by its user.

(ii) Quality Control

Quality control can be defined as the controls applied at each manufacturing stage to consistently produce a quality product.

(iii) Standardization

Standardization of a process can be defined as the setting up of process parameters so that it consistently produces a product of uniform characteristics.

(iv) Codes

Codes are conventionalized set of rules of good practice. A prime example of codes is the ASME Boiler and Pressure Vessel Code which is a set of standards that assure the safe design, construction and testing of boilers and pressure vessels. ASME, Section V, is for Non-destructive Testing covering methods such as Liquid Penetrant, Magnetic Particle, Radiography, Ultrasonic and Eddy Current Testing.

(v) Standards

Standards are documents that govern and guide the various activities occurring during the production of an industrial product. Standards describe the technical requirements for a material, process, product, system or service. They also indicate, as appropriate, the procedures, methods, equipment or tests to determine that the requirements have been met. For example ANSI, ISO, IS etc.

(vi) Specification

Specification is a detailed description of the criteria for the constituent, construction, appearance etc., of a material and service. In other words, specifications are documents for communication of the engineering, management and purchasing levels.

(vii) Guides and Recommended Practices

Guides and recommended practices are standards that are offered primarily as aids to the user. They use verbs such as "should" and "may" because their use is usually optional. However, if these documents are referenced by codes or contractual agreements, their use may become mandatory. If the codes or agreements contain non-mandatory sections or appendices, the use of referenced guides and recommended practices by them, are at the user's discretion.

(viii) Procedure

Procedure is a written document made on step by step basis which shows HOW to do it. The procedure should specifically present everything that the inspector or operator needs to know. Procedures are frequently developed in laboratory conditions on special test objects. Under such circumstances, procedures should also be tried under production conditions by production personnel before actual implementation.

The purpose or function of the Standards and/or Codes is to prescribe or guide a test method or procedure which could give correct, reliable and uniform assessment of the quality of the product, and ensure that the method is reproducible.

(ix) Technique

A technique is a specific way of utilizing a particular non-destructive testing method. Each technique is identified by at least one particular important variable from another technique within the method (Example: RT Method-X-Rays/Gamma Ray Techniques).

(x) Examination and Testing

Examination and testing are those quality control functions which are carried out during the fabrication of an industrial product by a qualified person who is an employee of the manufacturer.

(xi) Inspection

Inspections are the quality control functions which are carried out during the fabrication of an industrial product by an "authorized inspector".

(xii) Authorized Inspector

An authorized inspector is a person who is not an employee of the fabricator of an industrial product but who is properly qualified and has the authority to verify to his satisfaction that all examinations specified in the construction code of the product have been made to the requirements of the referencing section of the construction code.

(xiii) Report

A report of a non-destructive examination or of welding or testing is a document which includes all the necessary information required to be able to:

(i) take decisions on the acceptance of the defects revealed by the examination.

(ii) facilitate repairs of non-acceptable defects, and

(iii) permit the examination or testing to be repeated

(xiv) Records

Records are documents which will give, at any time in the future, the following information about a non-destructive testing examination, (i) the procedure used to carry out the examination, (ii) the data recording and data analyzing techniques used, and (iii) the results of the examination.

(xv) Instruction

Instruction is a detailed description for performing a specific activity. It can be classified into two ways:

a) Verbal Instruction

b) Written Instruction

Verbal instruction is normally done by higher level personnel in the form of supervision before and/ordering the execution of an activity. Written instructions are a collection of classified jobs under the constraint of a procedure.

8.1.3 Applications of Codes and Standards

Thus, a code or standard is a set of rules or guide lines to achieve a degree of quality aimed at safe use of a product. While the term code is used when statutory laws or requirements are met by a standard, the term standard is used to indicate the level or degree of quality. For example Heat Exchanger Standards by Tubular Exchanger Manufacturer's Association (TEMA) Class 'A', Class 'B' and Class "C' are different standards for various applications of Heat Exchangers. Hence for a given product one can have a different set of specifications or standards to meet different service requirements.

The selection and applicability of a Code or a Standard will depend on the nature of the Component and its Design indented life and operating Conditions. Proper selection of a Code becomes important. The minimum requirements of a particular code or standard may not satisfy the special needs of every user. Therefore, a user may find it necessary to invoke additional requirements to obtain required quality.

8.1.4 Introduction to NDT Codes and Standards

Non-Destructive Examination is very significant means to ensure reliability of equipment and structural components. Quite often a product being designed and manufactured or fabricated follows a certain reference Standard, Code, or Specification. In relation to NDT inspection, the purpose of Standards and Codes is to prescribe a test method of procedure which would give correct and uniform assessment of the quality of the product. Competent NDT personnel are required to assess the quality of the product. This can only be achieved through proper training and education in NDT practices. This is why the qualification and certification of NDT personnel is being standardized through- out the world by drafting of an ISO standard.

The use and the performance of non-destructive testing in industry depend very much on Standards. Standards are used to compare results, to calibrate equipment, to ensure uniform and reproducible results, and to help one to determine what is acceptable and what is not. Generally standards and codes are classified in the following manner:

(i) Standards and codes on the product. This standard will normally specify or recommend, acceptable methods of manufacture and fabrication, requirements for test and inspection and also the acceptance criteria for the test and inspection.

Examples of such standards or codes are:

(a) ASME Boiler and Pressure Vessels Code Sections I, II and III

(b) BS 5500 Specification for Unfired Fusion Welded Pressure Vessels

(c) AWS D 1.1 Structural Welding Code- Steel etc.

(ii) Standards and codes on the method or procedure of test or inspection. Some of these types of standards specify general testing methods or procedures while others specify testing methods for specific products, shapes or materials such as pipe welds, etc. Some will also specify types, materials and shapes of the image quality indicators and sometimes contrast meters as in the Japanese Standards. Generally this type of standard will specify the radiographic arrangement, criteria for film selection, intensifying screen, selection of radiation energy, quality or sensitivity level etc. Some examples of such standards or codes are:

(a) ASME Boiler and Pressure Vessels Code Section V.

(b) BS 2600,

(c) BS 2910.

(iii) Standards for collection of reference radiographs. Depending on the purpose, there are various collections of reference radiographs, such as those showing the correspondence, between designation and shape of defects, those classifying defects by their shape, size, distribution, etc. and some even classify or grade the degree of severity of the comparison of radiographs of cast products whose judgment of the results cannot be easily determined quantitatively. Examples of such standards include:

(a) IIW Collection of reference radiographs for welds
(b) ASTM E 186 Reference radiographs for heavy walled steel castings
(c) ASTM E 390 Reference radiographs for steel fusion welds, etc.

The standards mentioned above, though basically similar, differ in their approach. The difference is noticed in the design concept of the IQI and thus the sensitivity, and also on the source to film distance. Thus to radiographically evaluate the quality of the products, generally it is required to make use of or refer to all or part of those three types of standards.

8.1.5 Some Important International and National Standardizing Bodies for NDT

- International Standards Organisation (ISO)
- American Society for Mechanical Engineers (ASME)
- American National Standards Institute (ANSI)
- International Institute of Welding (IIW)
- American Welding Society (AWS)
- ASTM (American Society for Testing and Materials)
- British Standard (BS) - (British Standards Association)
- Bureau of Indian Standards (BIS)
- Canadian Standard (CSA) -(Canadian Standards Association)
- Deutsches Institute fuer Normung (DIN-German)
- Japanese Industrial Standard (JIS)

8.1.6 Codes and Standards in Radiography

8.1.6.1 International Organization for Standardization (ISO)

ISO promotes the development of standards to facilitate the international exchange of goods and services. It is comprised of the standards-writing bodies of more than 80 countries and has adopted or developed over 4,000 standards. Bureau of Indian Standards (BIS) is the designated Indian Representative to ISO. Table - 8.1.1 lists some of the ISO standards pertaining to radiography.

8.1.6.2 International Institute of Welding (IIW)

The list of standards and documents published by IIW which relates to radiographic testing is given in Table-8.1.2. In addition the IIW have also prepared the following collections of reference radiographs:

(i) Collection of reference radiographs of welds in steel (1962) (86 radiographs)
(ii) Collection of reference radiographs of welds in aluminium and aluminium alloys (1965) (51 radiographs)

Table 8.1.1 List of ISO Standards Relating to Radiographic Testing

Sl. No.	Standard No.	Title
1	ISO/DIS/5576:1982	Industrial radiology-Non-destructive testing Vocabulary (ISO/TC 135)
2	ISO/R 947:1969	Recommended practice for radiographic inspection of circumferential fusion welded butt joints in steel pipes up to 50mm (2 in) wall thickness (ISO/RC 44)
3	ISO/R 1027:1969	Radiographic image quality indicators- Principles and identification (ISO/TC 44)
4	ISO/R 1106:1969	Recommended practice for radiographic inspection of fusion-welded butt joints for steel plates up to 50mm (2") thick (ISO/RC 44)
5	ISO 1677:1977	Sealed radioactive Sources- General (ISO/TC 85)
6	ISO 1758:1976	Direct Reading electroscope –type pocket exposure meters (ISO/RC 85)
7	ISO 2405:1972	Recommended practice for radiographic inspection of fusion-welded butt joints for steel plates 50-200mm thick (ISO/TC 44)
8	ISO 2437:1972	Recommended practice for the X-ray inspection of fusion-welded butt joints for aluminum and its alloys and magnesium and its alloys 5-50mm thick (ISO/RC 44)
9	ISO 2504:1973	Radiography of welds and viewing conditions for film utilization of recommended patterns of image quality indicator (IQI) (ISO/RC 44)
10	ISO 2604:1975	Steel products for pressure purposes – Quality requirements (ISO/RC 17)
11	ISO 2919:1980	Sealed radioactive Sources – Classification (ISO/TC 85)
12	ISO 3777:1976	Recommended practice for the radiographic inspection of resistance spot welds for aluminium and its alloys (ISO/RC 44)
13	ISO 3897:1976	Storage conditions for silver image photographic plates for record purposes (ISO/TC 42)
14	ISO 3825:1978	Identification and certification of Unsealed radioactive source (ISO/.TC 85)
15	ISO 3999:1977(E)	Specification of apparatus for Gamma Radiography (ISO/TC 85)
16	ISO 3999-1:2000(E)	Radiation protection – Apparatus for industrial gamma radiography
17	ISO 5580:1985(E)	Non-destructive testing-Industrial radiographic illuminators-Minimum requirements
18	ISO 5655:2000 (E)	Photography-Industrial radiographic films (roll and sheet) and metal intensifying screens-Dimensions
19	11699-1:1998 (E)	Non-destructive testing-Industrial radiographic films
20	11699-2:1998 (E)	Non-destructive testing-Industrial radiographic films

Table 8.1.2 List of IIW Standards Related to Radiography

Sl. No.	Standard/Document Number	Title
1	IIS/IIW-183-65	Recommendation for the determination of the focal spot size of X-ray tubes
2	IIS/IIW-185-65	Rules for the reading of the IQI recommended by Commission V
3	IIS/IIW-208-66	Report on the investigation into the influence of plate thickness on the visibility of weld defects and on image quality in radiography of welds.
4	IIS/IIW-412-72	The use and limitations of radiographic image quality indicators.
5	IIS/IIW-423-73	Recommended practice for the radiographic examination of fusion-welded butt joints in steel plates from 0.5 to 50mm thick.
6	IIS/IIW-424-73	Image quality indicators for use in the radiography of aluminum and its alloys thickness up to 50 mm.
7	IIS/IIW-492-75	Recommended practice for radiographic inspection of fusion-welded circumferential joints in steel pipes from 1mm up to 50mm thickness (Revision of IIS/IIW-36-59)
8	IIS/IIW-585-79	Recommendations relating to the training of non- destructive testing personnel.
9	IIS/IIW-636-80	Inspection of welds when fitness-for-purpose criteria are applied preliminary recommendations.

(iii) Collection of reference radiographs and cross sections of welds in steel (34 radiographs and 34 drawings/cross sections)

8.1.6.3 American Society for Testing and Materials (ASTM)

ASTM develops and publishes specification for use in the production and testing of materials. It covers virtually all materials used in industry and commerce except welding consumables, which are covered by AWS specifications. It is currently composed of 15 sections comprising 65 volumes and an index. Section 1 covers iron and steel products; Section 2, non-ferrous metal products and Section 3, metal test methods and analytical procedures. Volume 3.03 in section 3 deals with the standard pertaining to non-destructive testing. The standards pertaining to radiography are listed in Table-8.1.3.

8.1.6.4 Bureau of Indian Standards (BIS)

BIS is the standards body in India responsible for the formulation of various standards. BIS is a member of the ISO. The various BIS standards in radiography are listed in Table –8.1.4.

Table 8.1.3 ASTM Standards on Radiography

Sl.No.	Standard No.	Title
1	E 94-00	Standard Guide for Radiographic Examination
2	E 142-92	Standard Method for Controlling Quality of Radiographic Testing
3	E 155-00	Standard Reference Radiographs for Inspection of Aluminium and Magnesium Casting
4	E 186-98	Standards Reference Radiographs for Heavy-Walled (2 to 4 $^{1/2}$ –in.(51 to 114-mm)) Steel Casting
5	E 192-95 (Reapproved 1999)	Standard Reference Radiographs for Investment Steel Casting of Aerospace Applications
6	E 242-01	Standard Reference Radiographs for Appearances of Radiographic Images as Certain Parameters are Changed
7	E 272-99	Standard Reference Radiographs for High-Strength Copper-Base and Nickel-Copper Alloy Castings
8	E 280-98	Standard Reference Radiographs for Heavy-Walled ($4^{1/2}$ to 12 in.) 114 to 305mm)) Steel Castings
9	E 310-99	Standards Reference Radiographs for Steel Fusion Welds
10	E 390-01	Standard Reference Radiographs for Steel Fusion Welds
11	E 431-96 (Reapproved 2002)	Standard Guide to Interpretation of Radiographs of Semiconductors and Related Devices
12	E 446-98	Standard Reference Radiographs for Steel Castings up to 2 in. (51 mm) in Thickness
13	E 505-01	Standard Reference Radiographs for Inspection of Aluminium Die Castings
14	E 545-99	Standard Test Method for Determining Image Quality in Direct Thermal Neutron Radiographic Examination
15	E 592-99	Standard Guide to Obtainable ASTM Equivalent Pentameter Sensitivity for Radiography of Steel Plates to 2 in. (6 to 51 mm) Thick with X-rays and 1 to 6 in. (25 to 152 mm) Thick with Cobals-60
16	E 689-95 (Reappoved 1999)	Standard Reference Radiographs for Ductile Iron Castings.
17	E 746-02	Standard Test Methods for Determining Relative Image Quality Response of Industrial Radiographic Film Systems.
18	E 747-97	Standard Practice for Design Manufacture and Material Grouping Classification of Wore Image Quality Indicators (IQI) Used for Radiology
19	E 748-95	Standard Practices or Thermal Neutron Radiography of Materials
20	E 801-01	Standard Practice for Controlling Quality of Radiological Examination of Electronic Devices

Table 8.1.3 contd.

Table 8.1.3 contd.

21	E 802-95 (Reapproved 1999)	Standard Reference Radiographs for Gray Iron Castings up to $4^{1/2}$ in. (114 mm) in Thickness
22	E 803-91	Standard Method for Determining the l/D Ratio of Neutron Radiography Beams
23	E 999-99	Standard Guide for Controlling the Quality of Industrial Radiographic Film Processing
24	E 1000-98	Standard Guide for Radioscopy
25	E 1025-95	Standard Practice for Design, Manufacture and Material Grouping Classification of Hole-Type Image Quality Indictors (KM) Used for Radiology
26	E 1030-00	Standard Test method for Radiographic Examination of Metallic Castings
27	E 1032-01	Standards Test Method for Radiographic Examination of Weldments
28	E 1114-92	Standard Test method for Determining the Focal Size of Iridium-192 Industrial Radiographic Sources
29	E 1161-95	Standard Test Method for Radiologic Testing of Semiconductors and Electronic Components
30	E 1165-92 (Reapproved 2002)	Standard Test Method for Measurement of Focal Spots of Industrial X-ray Tubes by Pinhole Imaging
31	E 1212-99	Standard Practice for Quality Control Systems for Non-Destructive Testing Agencies
32	E 1254-98	Standard Guide for Storage of Radiographs and Unexposed Industrial Radiographic Films
33	E 1255-96 (Reapproved 2002)	Standard Practice for Radioscopy
34	E 1320-00	Standard Reference Radiographs Titanium Castings
35	E 1390-90 (Reapproved 2000)	Standard Guide to Illuminators Used for Viewing Industrial Radiographs
36	E 1320-00	Standard Practice for Qualification of Radioscopic Systems
37	E 1416-96	Standard Test Method for Radioscopic Examination of Weldments
38	E 1453-98	Standard Guide for Storage of Media that Contains Analog or Digital Radioscopic Data
39	E 1475-97	Standard Guide in Data Fields for Computerized Transfer of Digital Radiological Test Data
40	E 1496-97	Standard Test Method for Neutron Radiographic Dimensional Measurement
41	E 1647-98	Standard Practice for Determining Contrast sensivity in Radioscopy
42	E 1648-95 (Reapproved 2001)	Standard Reference Radiographs for Examination of Aluminium Fusion Welds

Table 8.1.3 contd.

Table 8.1.3 contd.

43	E 1734-98	Standard practice for Radioscopic Examination of Castings
44	E 1735-95 (Reapproved 2000) el.	Standard Test method for Determining Relative Image Quality of Industrial Radiographic Film Exposed to X-Radiation from 4 to 25 Mv
45	E 1742-00	Standard Practice for Radiographic Examination
46	E 1815-96 (Reapproved 2001)	Standard Test Method for Classification of Film Systems for Industrial Radiography
47	E 1817-96	Standard Practice for Controlling Quality of Radiological Examination by using Representative Quality Indicators (RQIS)
48	E 1862-97 el.	Standard Test Methods for Measuring and Compensating in Reflected Temperature Using Infrared Imaging Radiometers
49	E 1933-999	Standard Test methods for Measuring and Compensating for Emissivity using Infrared Imaging Radiometers
50	E 1936-97	Standard Reference Radiographs for Evaluating the Performance of Radiographic Digitization systems
51	E 1955-98	Standard Radiographic Examination for Soundness of Welds In Steel by Comparison to Graded ASTM E 390 Reference Radiographs
52	E 2003M-98	Standard Practice for Fabrication of the Neutron Radiographic Beam Purity Indicators
53	E 2003-98 el	Standard Practice for Fabrication of the Neutron Radiographic Beam Purity Indicators
54	E 2007-00	Standard Guide in Computed Radiology (Photostimulable Luminescence (PSL) Method)
55	E 2023-99	Standard practice for Fabrication of Neutron Radiographic Sensitivity Indicators
56	E 2104-01	Standard Practice for Radiographic Examination of Advanced Aero and Turbine Materials and Components

Table 8.1.4 Typical IS Standards on Radiography

Standard number	Title
IS 2598 - 1966	Safety code for Industrial radiography practices.
IS 2595 - 1693	Code practice for radiography testing.
IS 1182 - 1967	Recommended practice for the radiographic examination of fusion welded butt joints in steel plates.
IS 3657 - 1966	Specifications for image quality indicators.
IS 4853 - 1968	Recommended practice for radiographic examination of fusion welded circumferential joints in steel pipes.

SECTION 9

RADIATION SAFETY

After reading this section you should be able to answer

★ *What are the biological effects of Radiation?*

★ *Define the units used in radioactivity and radiation monitoring.*

★ *Identify appropriate methods for monitoring the radiation levels in and around radiographic installations and also the dose received by radiographic workers.*

★ *What are the safety precautions to be observed while transporting radioactive material.*

Chapter 9.1
Biological Effects of Radiation

The early experimenters with radiations (primarily X-rays) were largely unaware of the hazards of radiation and were also not very much worried about personal safety. The major side effect of irradiation commonly observed was the reddening of exposed skin – erythema. In fact some radiographers even used this erythema to gauge X-ray beam strength. In 1904, the first X-ray fatality in US was reported. A dramatic demonstration of radiation hazard occurred in the 1920s when luminous watch dials came into production. Workers who painted the numbers on the watch dials followed their customary practice of dipping brushes in the paint and then twirling the tips with their lips to make them pointed. The small amount of radium in the luminous paint got into their bodies and lodged themselves in the bones. Years later, after they had stopped working, these personnel began to die mysteriously. Investigations revealed that their death was due to bone cancer. The atomic bombing of Hiroshima and Nagasaki demonstrated the havoc which radiation can play on man. Thus, over the years, radiographers, radiologists, radium dial painters, uranium miners, atomic bomb survivors and many others have provided tragic and extensive evidence of the hazards of ionising radiation. Thus, while radiation and radioactivity has made it possible to diagnose and cure diseases, ensure the safety and reliability of engineering components and plants, it has also damaged many a lives. It is thus imperative that to derive the maximum benefits from these, the hazards associated with radiation are well understood so that adequate precautions can be taken. This is specially true in the case of industrial radiographers who are known to be among the highly exposed group of personnel who are occupationally exposed to ionising radiations.

Human beings receive radiation from natural and artificial sources. Natural sources include cosmic rays, naturally occurring radioisotopes such as uranium, thorium, etc. Artificial sources include electronic devices (TV), radioactivity released on account of atmospheric testing of nuclear devices, X-ray machines, radioisotopes, etc. Radiations can induce detrimental biological effects in organs and tissues by producing ions and depositing energy that may cause damage. In this chapter, we briefly discuss the biological effects of radiation.

> All of us receive radiation both from natural and artificial sources.
>
> Radiations can induce detrimental biological effects.
>
> Basic building blocks of tissues and organs are cells.
>
> Cells contain nucleus and cytoplasm.
>
> Nucleus directs the cellular activity while these activities occur in the cytoplasm.

9.1.1 Structure of the Cell

The biological effects of ionising radiations are primarily due to the interaction of radiation with the cells in the body. The cell is the basic or the fundamental building block of any living species. All the tissues and organs are made up of cells. It is estimated that an adult human body consists of about 10^{14} cells. The sizes, type and shape of the cells vary from tissue to tissue and species to species. In the human body, the smallest cell is the leucocyte (WBC) having a diameter of 3-4 microns. Though different types of cells exist, all cells have a centrally located "nucleus" surrounded by the "cytoplasm". The nucleus controls the cell functions and directs the cellular activity. The cytoplasm is nothing but an aqueous solution of soluble proteins and salts. It is in the cytoplasm that cellular activity occurs. This activity includes synthesis of proteins, storage of energy and its release when needed. The cells originate and multiply from pre-existing cells by the process of cell division.

The nucleus consists of chromosomes which generally occur in pairs (except in cells called germ cells which are responsible for reproduction). The chromosome contains the genes, which characterise the genetic or hereditary features.

9.1.2 Radiation Effects on Cells

The radiation effects on cells can be explained in terms of direct action and indirect action. In the direct action, the sensitive volume in the cell is changed/inactivated by the direct absorption of energy from the radiation in question. In the indirect action, the sensitive volume is inactivated by transfer of energy from another volume, which has directly or indirectly absorbed energy from radiation. In other words, while in direct action, the volume which absorbs energy is the volume which is inactivated, in indirect action, the volume which is inactivated is not the volume which absorbs the energy. Natural mechanisms are available in the body, which can effectively repair the damage caused by radiation. When these mechanisms are not successful, irradiation of cells can lead to the following effects

(a) *Stoppage of Cell Division:* In the human body some types of cells are always in the process of division to replace worn out cells. Inhibition of mitosis refers to the delay in the process of cell division. The effect can be temporary or permanent depending on the severity of the radiation dose and dose rate.

(b) *Chromosomal aberration:* Radiation can cause breaks in chromosomes. This may result in some part of the genetic material not being transferred to daughter cell. However, a majority of breaks may get repaired and the damage may not manifest. The number of breaks can be correlated to dose. This effect serves as a biological indicator of radiation exposure and can be employed for dosimetry in case of radiation accidents to ascertain the genuineness of radiation exposure.

(c) *Gene mutation:* This is the change in the structure of the gene and hence in the characteristic of the gene which will manifest as genetic effects.

(d) *Cell death:* Irradiation can result in the death of the cell due to changes in the physical properties of vital cell structure.

> Nucleus contains Chromosomes.
> Chromosomes contain genes.
>
> Genes carry the hereditary features.
>
> Exposure of cells to radiation can lead to (depending on dose and period)
> * *Stoppage of cell division*
> * *Chromosomal aberration*
> * *Gene mutation*
> * *Cell death*

Lymphocytes, oocytes, and spermatogonia are killed by a few Sieverts while cells in organs where the rate of cell division is low require several ten to hundred Sieverts to cause its death.

9.1.3 Biological Effects

Biological effects of radiation can be of two types :

(a) Somatic effects – These effects manifest in the individual exposed to radiation.

(b) Genetic effects – These are effects which manifest in the future generations of persons exposed to radiation. These are also called as hereditary effects.

The biological effects may result from two types of exposures namely acute exposure (high exposure in short periods) and chronic exposure (small exposure over long period). The biological effects may be classified as immediate effects (those which manifest within a few weeks of exposure) and delayed effects (those which manifest after a few years). These are listed in Table - 9.1.1.

Table 9.1.1 Biological Effects

Immediate Effects (due to acute exposure)	Delayed Effects (due to chronic or acute exposure)
Chromosome aberration	Cancer
Blood Changes	Leukaemia
Nausea, vomiting, diarrhea (NVD)	Cataract
Radiation sickness	Hereditary effects
Epilation	
Skin erythema	
Sterlity	
Death	

9.1.3.1 Lethal Dose

The radiation sensitivity or response varies considerably from one species to another and to compare the radiation responses, the term lethal dose ($LD_{50/30}$) is used. This is the whole body acute dose required to kill 50% of the exposed organisms within 30 days after irradiation. The $LD_{50/30}$ for man is about 4 Sv -6 Sv

> Biological effects can be classified as
>
> *Somatic effects manifest in the individuals exposed to radiation*
>
> *Genetic effect manifests in future generations.*
>
> Severity of the effect depends on radiation dose
>
> $LD_{50/30}$ for humans is 4 Sv-6 Sv

9.1.3.2 Radiation Sickness

Human beings are quite sensitive to radiation. Symptoms of radiation sickness, i.e. sickness arising as a result of irradiation (also known as radiation syndrome) are summarised in Table- 9.1.2. The effects will vary depending on the organs, age, food habit etc. An important and unfortunate feature is that except for vomiting sensation which sets in after an hour or so after a whole body dose above 2 – 3 Gy (200 – 300 rads), the victim has no indication that he has received a fatal or near fatal dose.

Table 9.1.2 Biological Effects due to Acute Dose

(A) Whole Body Irradiation

Dose Range	Immediate Effect
<100 mSv (<10 rem)	None detectable
0.1 Sv to 0.25 Sv (10 rem to 25 rem)	Chromosome aberration, Recoverable
0.25 Sv to 1 Sv (25 rem to 100 rem)	Change in blood picture, Recoverable
1 Sv to 3 Sv (100 rem to 300 rem)	Nausea, Vomiting, Diarrhea (NVD), Loss of appetite, Recoverable probable
3 Sv to 5 Sv (300 rem to 500 rem)	NVD + Radiation fever, Recoverable probable
>5 Sv (500 rem)	Death within few days to weeks
4 Sv to 6 Sv (400 – 600 rem)	$LD_{50/30}$
10 Sv (1000 rems)	NVD in all within two hours
50 Sv (5000 rems)	Incapacitation immediately, Death within few hrs to 1 week

(B) Local Irradiation

Dose	Region	Effect
0.25 Sv (25 rem)	Gonads of men	Temporary sterility
3 Sv (300 rem)	Skin	Epilation
1.6 Sv (160 rem)	Gonads of Women	Temporary sterility
3.2 Sv (320 rem)	Gonads of Women	Permanent sterility
6 Sv (600 rem)	Skin	Skin erthema in 30% exposed persons
8 Sv (800 rem)	Eye lens	Cataract after 5-10 yrs
9.6 Sv (960 rem)	Gonads of men	Permanent sterility

9.1.3.3 Stochastic and Non Stochastic Effects

For certain biological effects, the probability of the effect depends upon the dose. These are called stochastic effects. For certain biological effects, the severity of the effect increases with the dose; for these, a threshold dose may be defined below which the effects will not occur. Such biological effects are called non-stochastic effects. Examples of stochastic effects are induction of cancer, leukaemia and hereditary effects. The hereditary effects arise due to gene mutations in the parent which are transferred to the future generations. Gene mutations have been occurring due to natural causes. Radiation also induces gene mutations. Examples of non-stochastic effects are NVD syndrome, death, epilation, skin erythema, cataract and sterility.

A great deal of information has been accumulated over the past 50 years about radiation exposures and the possible effects. This information has come from animal studies and also from studies of different groups of people who had received quite large exposures eg., medical radiologists exposed to X rays. Uranium miners and luminous dial painters exposed to radium and radon, patients exposed to X rays for treatment of certain diseases, survivors of Hiroshima and Nagasaki and others.

Based on the analysis of the effect of radiation on these groups of people, the ICRP has arrived at certain dose limits that can be considered quite safe. That is to say, if work is within these dose limits as recommended by ICRP, the probability for appearance of any of the above biological effects is negligible. These limits are given for uniform exposure. In the case of partial body exposures i.e., exposures to individual organs, the limits are somewhat higher. From the point of view of radiation protection, the points to remember are:

a. most of the non-stochastic effects can be prevented if the acute doses are below 500 mSv (50 rem)
b. doses received at low dose rates do not result in non-stochastic (immediate) effects.
c. probability of the stochastic effect is very small if the total cumulative dose received by a person (either by acute or by chronic exposure) is kept low.

Chapter 9.2
Radiation Units and Operational Limits

9.2.1 Radiation Units

To express radiation quantitatively, personnel involved in industrial radiography should have adequate knowledge of the various units of radiation measurement. The term dose is generally used to indicate the radiation exposure received by a person or an object. Basically two concepts are used to define the various terms of radiation measurement – the ionisation of matter by radiation and the energy absorbed by objects (living and non living) from the impinging radiation. These two concepts can be subdivided into four main groups namely, (1) Activity (2) Exposure (3) Dose and (4) Dose rate.

9.2.1.1 Activity

Radioactivity is the property of some isotopes to spontaneously emit one or more of the alpha or beta radiations with or without gamma radiations. When a radioactive nucleus changes to another nucleus, the change is called decay or disintegration. Quantitatively, the rate of decay or activity is expressed in terms of the number of disintegrations per unit of time (usually seconds or minutes). The unit of activity is Becquerrel (Bq). One Bq corresponds to one disintegration per second. The older unit of activity - Curies (Ci) is till widely used. One Curie corresponds to 3.7 x 10^{10} disintegrations per second.

$$1 \text{ Ci} = 3.7 \times 10^{10} \text{ Bq} \qquad \text{... (9.2.1)}$$

9.2.1.2 Exposure

When radiation passes through air it causes ionisation. Exposure is a quantity expressing the amount of ionisation caused in air by the passage of X- or gamma radiation. The unit of exposure is coulomb per kg. Roentgen is the older unit of exposure but still widely used by radiographers. Roentgen is defined as the amount of X- or gamma radiation that produces one electrostatic unit (esu) of charge of either sign as a result of interactions in 1 cc of dry air at standard temperature and pressure (STP). One Roentgen corresponds to an energy absorption of 86.9 ergs per gram of air.

$$1 \text{ R} = 2.58 \times 10^{-4} \text{ coulombs/kg.} \qquad \text{... (9.2.2)}$$

> Activity – rate of decay
> * *Unit - Becquerel*
> * *Older unit Cuire*
> * *1Ci = 3.7x10^{10}dps*
> * *1 Bq= 1dps*
>
> Exposure – amount of ionisation caused in air by passage of X- or gamma rays
> * *Unit –Coulomb/kg*
> *Older unit - Roentgen still most widely used*

Important points to be noted here are that

(a) the exposure unit is the total exposure and does not involve the time over which the exposure is made and

(b) the unit Roentgen is valid only for (i) air as a medium and (b) X- and gamma rays of energy less than 3 MeV.

9.2.1.3 Dose

Dose is a measure of the amount of energy deposited in the matter (any medium) by the ionising radiation per unit mass of irradiated material at the place of interest. The unit of dose is Gray (Gy). The older unit of dose is "rad" which is also still in use.

$$100 \text{ rad} = 1 \text{ Gray (Gy)}. \qquad \text{... (9.2.3)}$$

The Gray is the unit for absorbed dose and represents an energy absorption of 1 Joule/kg of any medium. The rad on the other hand represents the absorption of 0.01 Joule/kg i.e., 100 ergs per gram. The unit Gray (or rad) is valid (a) for all types of radiation (b) for any energy and (c) in any medium.

9.2.1.4 Dose Equivalent

Same amount of different types of radiation have different effects in biological materials like human beings. Thus if 1 gray of X-rays and 1 gray of alpha particles impinge on a human body, the alpha particles would cause greater damage compared to X-rays. This is because ionisation produced by alpha radiation is more intense compared to ionisation produced by X- or gamma rays. The parameter known as equivalent dose is, therefore, defined to take into account the quality of radiation. Dose equivalent is a quantity used for purpose of radiation protection to express the effect of all ionising radiations on a common scale. The unit of dose equivalent is Sievert (Sv). The older unit which is still widely used is the rem (roentgen equivalent man).

$$1 \text{ Sievert (Sv)} = 100 \text{ rem}. \qquad \text{... (9.2.4)}$$

Dose equivalent (Sv)

$$= \text{radiation weighting factor } (W_R) \text{ x absorbed dose (Gy)} \qquad \text{... (9.2.5)}$$

The radiation weighting factor also known as the quality factor is defined as the ratio of the absorbed dose of X- or gamma rays to the absorbed dose of any radiation required to produce the same effect. Smaller units of dose equivalent such as milli Sievert (mSv) or milli rem (mrem) are more often used.

$$1 \text{ Sv} = 1000 \text{ mSv and } 1 \text{ Rem} = 1000 \text{ mRem} \qquad \text{... (9.2.6)}$$

Table – 9.2.1 lists the radiation weighting factors for different types of radiation and the energy range.

> Dose – measure of energy deposited in a medium by any type of radiation
> * *Unit – Gray (Gy)*
> * *Older unit rad*
> * *1Gy = 100 rad*
>
> Dose Equivalent – radiation dose taking into account quality of radiation.
> * *Unit – Sievert*
> * *Older unit - rem still most widely used*
> * *100 rem = 1 Sv*

Table 9.2.1 Weighting Factors for various types of radiation

Radiation type and energy range	Radiation Weighting Factor (W_T)
X- and Gamma rays all energies	1
Beta particles - all energies	1
Neutrons, energy	
< 10 keV	5
> 10 keV to 100 keV	10
> 100 keV to 2 MeV	20
> 2 MeV to 20 MeV	10
> 20 MeV	5
Protons, other than recoil protons, energy > 2 MeV	5
Alpha particles, fission fragments, heavy nuclei	20

9.2.1.5 Dose Rate

Dose rate is the dose delivered or received per unit time. The 'hour' is commonly taken as the unit of time here (Example: R/hr, mR/hr; rad/hr, Gy/hr; Sv/hr, rem/hr, mSv/hr, mrem/hr). Consider a person working in a radiation area having a dose rate of 60 mR/hr (0.6 mGy/h). If he works for duration of 10 minutes, he will receive a dose of 10 mR (0.1 mGy). Strictly we should speak of it as 10 mrad (0.1 mGy) absorbed dose or 100 micro Sievert dose equivalent. If the person works in the same field for 4 hours, he will receive 40 mrad (0.4 mGy) of absorbed dose.

The various units (new and old) are summarised in Table – 9.2.2 below.

Table 9.2.2 Units at a glance

Sl. No.		Old Unit (Abbreviation)	SI unit (Abbreviation)	Relationship
1.	Activity	Curie (Ci)	Becquerel (Bq)	1 Ci = 3.7×10^{10} Bq
2.	Exposure	Roentgen	Coulomb/kg (C/kg)	1 R = 2.58×10^{-4} C/kg
3.	Absorbed Dose	Rad	Gray (Gy)	100 Rad = 1 Gy
4.	Biological Dose	Rem	Sievert (Sv)	100 Rem = 1 Sv

9.2.2 Operational Limits

Since exposure to radiation is harmful, the ideal situation would be no exposure to radiation at all. However, this is not possible due to the immense benefits of radiation and radioactive materials. Mankind is going to use radiation and radioactive materials not only now but also in the years to come. Hence, regulations are needed to allow their use within safe limits. This is the issue, which has been addressed by the International Commission of Radiological Protection (ICRP). Radiation protection is concerned with the protection of

(a) individuals,
(b) their future generation and
(c) mankind as a whole,

in the course of the use of radiation for beneficial purposes. The aim of the rules and regulations issued for radiation protection should be to

(a) make certain that work involving radiation exposure is necessary and justified.
(b) prevent non-stochastic effects and
(c) limit the probability of stochastic effects to an acceptable level consequently, the ICRP recommended a system of dose limitation based on

Justification: No exposure to ionising radiation shall be permitted unless its introduction results in a net positive benefit.

Optimisation: A sustained effort should be made to ensure that "All doses received are maintained As Low As Reasonably Achievable (ALARA)" economic and social factors being taken into account and

Limitation: Dose equivalents received during a year or accumulated over a period of time shall not exceed the limits recommended by the Commission.

A basis for the ICRP recommendation is the assumption that any exposure to radiation may carry some risk of somatic and hereditary effects. It is therefore necessary to limit the dose to a level at which the assumed risk is deemed to be acceptable to the individuals and to the society in view of the benefits derived from such activities.

ICRP was founded in 1928 and issues periodic recommendations on radiation protection. The ICRP Publication 60 "1990 recommendations of the International Commission of Radiological Protection" (1991) is the basis of most current regulations in India and many other countries. The Atomic Energy Regulatory Board (AERB) in India is the competent authority in India as regards radiation, radioactivity and handling of radiation sources is concerned.

9.2.2.1 Dose Equivalent Limit

The dose equivalent limit was earlier known as Maximum Permissible Dose (MPD). For an individual, it is that dose accumulated over a long period of time or resulting from a single exposure, which in the light of the present knowledge carries a negligible probability of severe somatic or hereditary injuries.

For recommending dose equivalent dose equivalent limits individuals are categorised into two groups.

(a) Adults exposed in the course of their work
(b) Members of the public

Persons belonging to the second category (including children) do not make their own choice to be exposed and may receive no direct benefit from the exposure. Therefore, for planning

> ICRP recommends ALARA
>
> *All doses received are maintained As Low As Reasonably Achievable (ALARA) economic and social factors being taken into account*

purposes, it is considered appropriate to set the dose equivalent limits for members of the public by a factor of twenty less than those for radiation workers.

A person can receive radiation doses from external and/or internal sources. Radionuclides can get into the body through ingestion, inhalation, wounds or absorption by skin. The dose equivalent limits are same whether the exposure is due to external sources, internal sources or combination of the two. Accumulation of internally deposited radionuclide will depend on the concentration of radionuclide in air and water and their (air & water) intake. Limits are therefore prescribed in terms of Annual Limits of Intake (ALI) for different radionuclides. Table – 9.2.3 summarises the maximum permissible level of exposures as recommended by ICRP.

Maximum Permissible Dose (MPD)

For occupational workers- 20 mSv /year

For general public – 1 mSv / year

MPD

includes doses due to external and internal exposure

does not include medical exposures.

Table 9.2.3 Summary of Dose Limits Recommended by ICRP

Sl. No.		Occupational	Public
1.	Whole body (Effective dose)	20 mSv/year (2 rem/year) averaged over 5 consecutive years 30 mSv in any single year	1 mSv/year (0.1 rem/year)
2.	Lens of the eyes (Equivalent dose)	150 mSv (15 rem) in a year	15 mSv (1.5 rem) in a year
3.	Skin, hands and feet (Equivalent dose)	500 mSv (50 rem) in a year	50 mSv (5 rem) in a year

The dose limits do not apply for medical exposures since the exposure is only for the benefit of the patient. However, the exposures received by the medical and paramedical staff are considered as occupational exposures and the dose limits apply for these exposures.

Chapter 9.3
Radiation Monitoring and Control

Radiography personnel are exposed to penetrating radiations while working with radiation sources. Unlike heat or light radiation, human body cannot feel the presence of ionizing radiations. Unless, the exposure to ionising radiation is extremely high and also occurs in a short time, radiation exposures would go unnoticed. As mentioned in Chapters 9.1 and 9.2, radiation exposures can prove to be hazardous. Hence, detection and monitoring of radiation levels in working areas by appropriate instruments is very essential to protect the personnel from undue exposure. Such monitoring would also aid in devise methods and procedures to control the exposures. Thus, radiation monitoring forms an important part of work practice. All methods of radiation detection and monitoring are based on the ability of radiation to cause ionisation Various radiation measuring devices differ in the medium of ionisation and the method of this ionisation. Radiation monitoring can be broadly classified into a) area monitoring and b) personnel monitoring.

9.3.1 Area Monitoring

The assessment of radiation levels at different locations in the vicinity of radiation sources is generally known as area monitoring. These measurements will give an idea about the radiation safety status of the installation and also confirm the adequacy of the existing radiation protection measures.

Area monitoring is essential

a. during field radiography operations with gamma sources to mark and cordon off areas with high radiation levels so as to minimise exposure of personnel and public.

b. assess the radiation levels around the gamma radiography cameras immediately after retraction of the cable drive to ensure that the source has returned back into the camera and is lodged safely.

c. assess the radiation levels around cameras or transport containers to ensure the adequacy of shielding. the maximum permissible leakage radiation levels around the camera with the beam in off condition are given in Table – 9.3.1. the radiation levels around the transport container is given in Table – 9.3.2.

d. to locate the source in case it gets stuck in the guide tube or is lost during transit so that immediate measures for shielding/retrieval can be made

> Radiation Monitoring done through Personnel Monitoring and Area Monitoring
>
> Area Monitoring refers to assessment of radiation levels in the vicinity of radiation sources.

e. assess radiation levels around the control panel of X-ray machine or around the installations during radiography to ensure that the radiation levels are within the permissible limits.

f. around the X-ray machine (other than in the primary x-ray beam direction) to ensure that the leakage radiation is within limits. The maximum permissible leakage radiation levels around the X-ray machines in given in Table – 9.3.3.

Table 9.3.1 Permissible leakage level around radiography cameras

Radiation measurement location	Average		Maximum	
	mGy/hr	mR/hr	mGy/hr	mR/hr
5 cm from the surface	0.2	20	1.0	100
1m from the source position	0.02	2	0.1	10

Table 9.3.2 Permissible leakage level around transport containers

Radiation measurement location	mGy/h	mR/h
On the surface	2	200
At 1m from surface	0.1	10

Table 9.3.3 Permissible leakage level through X-ray machines

The maximum leakage radiation level at 1 metre from the tungsten target should not exceed 1 R in 1 hour.
The leakage radiation level through the tube housing is measured in the ON position of the X-ray beam at its maximum ratings (maximum kV and maximum mA), in directions other than that of the primary beam

Considering the above aspects, it may be necessary to monitor the radiation levels ranging from as low as 0.25 mR/hr along the cordoned area to more than 10 R/hr in the case of stuck source. In case of high radiation fields, the radiation worker has to keep himself at a safe distance, while monitoring the location of interest. The monitoring instrument should have proper provisions for this purpose. These varying requirements necessitate the use of more than one type of instrument.

A variety of monitoring instruments are available each with a specific capability. Practically all the survey instruments employ an ionization chamber or a Geiger Muller (GM) counter or an inorganic crystal such as NaI (Tl) as detector. The following radiation survey instruments are generally being used for area monitoring:

(a) Gun monitor (ECIL type SM 140) or Beta gamma exposure rate meter,

(b) Mini Rad survey meter,

(c) Radiation survey meter (ECIL type MR 121C), and

(d) Teletector or wide range survey meter.

9.3.1.1 Gun monitor

It has an ionisation chamber detector and can measure dose rates from 5 mR/hr to 5 R/hr in three ranges. When the front cover of the detector is removed this instrument can be used to monitor beta radiation also. This instrument is useful for general area monitoring and for checking radiation levels around shielded source containers. This type of instrument is not useful for monitoring very low levels of radiation.

9.3.1.2 Mini Rad Survey Meter

Also known as "Mini Rad", this GM detector based survey meter is extensively used to monitor low as well as high radiation levels. This has 4 ranges viz., 5 mR/h, 50 mR/h, 500 mR/h and 5 R/h. This is a miniaturized, light weight and rugged survey meter ideal for field applications. Fig. 9.3.1 shows a typical minirad radiation survey meter.

Fig. 9.3.1 Minirad radiation survey meter
(Courtesy M/s Pulse Echo Systems, Mumbai, India)

9.3.1.3 Radiation Survey Meter (MR-121 C)

This instrument uses a GM counter and is popularly known as GM survey meter. It can be used to monitor radiation level from as low as 0.1 mR/h to 20 mR/h. It has three ranges viz., 0.2 mR/h, 2 mR/h and 20 mR/h. It is very useful for monitoring around the cordoned area during field operations and around radiography installations. The instrument is highly sensitive to low levels of radiation but fails at high radiation levels.

9.3.1.4 Teletector or Wide Range Survey Meter

Small GM counters are used as detectors and help to monitor very high radiation levels [1000 R/h]. This can be used for monitoring around radiography cameras, for determining radiation output of radiography sources and for locating a source stuck in the guide tube. The main advantage of this monitor is that radiation levels can be measured by standing at a distance from the source, thereby reducing radiation exposures to the person who is measuring.

In all these instruments the radiation level is indicated on a meter. In some instruments, in addition to the visual display and audio alarm is also provided. Such instruments give audible sound in the presence of radiation. This is a useful tool to alert the radiographer to the presence of radiation.

The use of monitors should be a part of the daily radiation work. It is essential to monitor in and around radiography installations during and after each exposure to confirm that the work procedure is proper and safe and that the system is functioning properly. It is essential for the radiography personnel to be familiar with the interpretation of the indicated reading and the capabilities of the equipment.

In the case of radiography rooms and source storage rooms, it is advisable to install fixed type of radiation zone monitors for continuous monitoring and recording of the levels of radiation. Many zone monitors are also provided with audio alarms which set off once the radiation level crosses a preset threshold.

9.3.2 Maintenance and Upkeep of Radiation Measuring Instruments and Devices

- All radiation measuring instruments should always be kept in good working condition and checked/calibrated periodically, once in six months and immediately after repairs.
- All instruments should always be stored in dust-free, dry and preferably air conditioned rooms. Toxic chemicals should not be stored near the instruments.
- The instruments should not be kept in a high radiation area or a contaminated area.
- *Care should also be taken to ensure that the instruments are not contaminated during their use.*
- When not in use for a long time, instruments operated on dry battery cells, should be stored after removing the batteries.
- The calibration certificates, and instructions or operating manuals of each instrument must be readily available in the instrument storage room.
- *Periodic checks on instrument's performance and calibration consistency should be made and relevant observations should be recorded and maintained in a log book kept along with the instruments.*
- *The most simple method of checking the instrument performance is to use the instrument just after it has been calibrated and recording for future reference the dose rate at a specific distance form a source of known strength. Performance checks can then be made at any time by comparison of the recorded reading with a check reading made at the same distance from the source. It may be necessary to make necessary correction for radioactive decay of the source. If the checked reading after applying the necessary corrections varies by ± 20%, the instrument should be sent for servicing or recalibration.*
- *All instruments should be duly recalibrated and up to date calibration factors should be obtained and used.*
- All defective instruments should be repaired without delay and recalibrated after repair.

◆ All personnel monitoring badges and dosimeters should be stored in a radiation free place. The personnel monitoring badges must be periodically returned to the designated authority for evaluation at predetermined intervals whether used or unused.

> Personnel Monitoring is for recording the dose received by individual radiation workers.
>
> Personnel monitoring done through film badge, pocket dosimeter and thermoluminescent dosimeter.
>
> Film badges uses radiation sensitive film with filters.
>
> Degree of film blackening is correlated to dose received.
>
> Filters in the badge serve to reduce energy dependence of radiation
>
> Film badge provides
>
> * *a permamnent record*
>
> * *Can record doses received from 20 mrem to 20 Rem.*

9.3.3 Personnel Monitoring

Personnel monitoring is the evaluation of the radiation doses received by the personnel working with radiation. The most commonly used devices for this purpose include film badge, thermoluminescent dosimeter and pocket dosimeter.

9.3.3.1 Film Badge Dosimeter

The film badge consists of a metal or a plastic holder containing one or more radiation sensitive films. The holder is generally provided with different filters (lead, cadmium, etc) and an open window. It is well known that the sensitivity of films vary with energy of radiation. The filters serve to reduce the energy dependence and to indicate the type and approximate quality of radiation. The film when exposed to radiation and processed under standard conditions of temperature and humidity, gets blackened just like ordinary films exposed to light. The degree of blackening depends on the radiation dose. Higher the dose, higher is the blackening or density. The density on the film is measured using a densitometer and correlated to the radiation dose using appropriate calibration graphs. Thus the dose received by the individual who wore the badge can be estimated.

Film badge is a routine monitoring device and should be worn by all radiation workers. In case of local irradiation such as hands, wrist badges are available, and this can be worn. A PM film packet once loaded in the cassette should be removed only when it is to be replaced with a new one. A control badge is normally provided with the film badges to correct for the background radiation and the climatic conditions at a particular site. Film badges when not in use along with the control badge should be kept at a safe location away from the place of work and where the radiation level is near to background level.

Advantages

a. It provides a permanent record of the exposure received
b. It gives information about the quality of radiation
c. It has a wide range and can record doses from a few mR to more than 10 R
d. It is inexpensive

Disadvantages

a. It has to be treated with reasonable care and should not be subjected to high temperature

b. It cannot give an immediate indication of the dose received during an accident

Pocket dosimeters - based on principle of gold leaf electroscope and the fact that radiation causes ionisation in air.

Provide immediately the dose received by the personnel during that period of work.

Available in various ranges. Most widely used range : 0 – 200 mrem

Must be charged to zero before use.

9.3.3.2 Pocket dosimeters

The pocket dosimeter is a small device approximately the size of a fountain pen and is based on two principles (a) like charges repel each other as in a gold leaf electroscope and (b) radiation causes ionisation in air.

It essentially consists of a quartz fiber and a wire each bent into a 'U' and joined at the ends to form a conducting system. A built-in microscope with a transparent scale is focussed on to the center of the bend in the quartz fiber (fig. 9.3.2). All these are housed in a metal cylinder. The electrode and the cylinder together form an electroscope. When an electrical potential is applied using a special charging unit, the quartz fibre and wire get charged with same type of charge and hence repel each other. When the dosimeter is properly charged, the indicator shall be at zero scale (as indicated in fig. 9.3.3) and the dosimeter is ready for use.

When the dosimeter is placed in a radiation area, the radiation causes ionisation of the air inside the chamber. As the charge on the electrode becomes neutral, the force of repulsion decreases resulting in the movement of the quartz fiber. This movement can be read on the scale and is a direct measure of the radiation dose received by the personnel during the period of time.

The dosimeter has a range of 0-200 mrem and it can be used over and over again after recharging. Other ranges cover 0-500 mrem, 0-5 R, 0-50 rem.

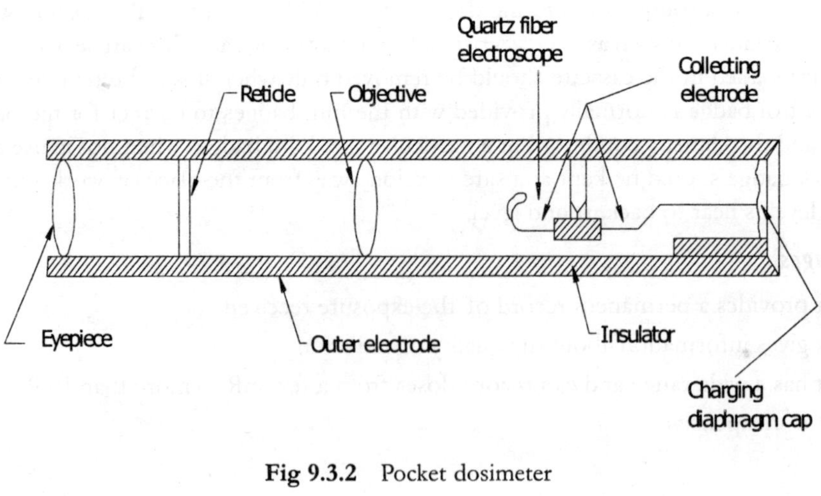

Fig 9.3.2 Pocket dosimeter

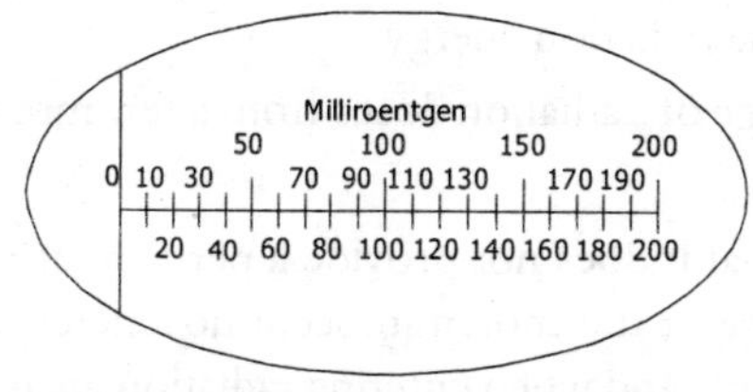

(a) Dosimeter fully charged. The quartz fiber rests near "O" on the scale

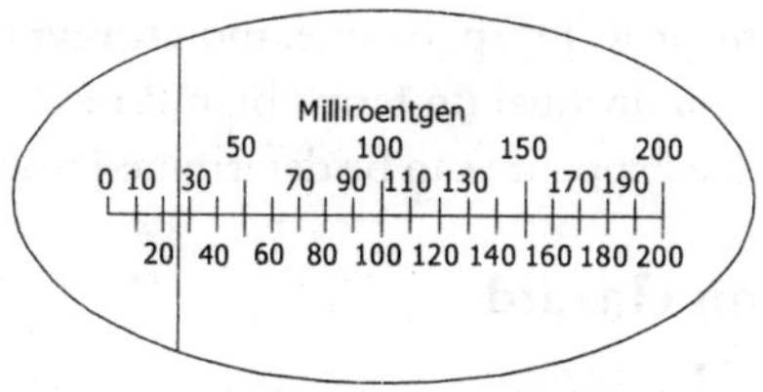

(b) Dosimeter exposed to radiation. The total dose received is recorded

Fig 9.3.3 Image viewed in self reading pocket dosimeter

Advantages

(a) It is fairly robust and does not need too much care.

(b) It is small and can be clipped easily onto clothing on any part of the body.

(c) It gives instantaneously the dose received by the personnel immediately after an exposure.

Disadvantage

(a) It does not provide a permanent record.

(b) It has a very limited range.

(c) It has to be charged properly before every use.

(d) It can give false readings under many circumstances: namely, being dropped on a hard floor, or due to electrical leakage that discharges the dosimeter.

9.3.3.3 Thermoluminescent dosimeter

This is a relatively new method of measuring radiation dose. The principle of this device is embodied in its name, "thermo" meaning "heat" and "luminescent" meaning "light producing". When thermoluminescent materials such as Lithium Fluoride (LiF) are exposed to radiation, electrons in these atoms are raised from the ground state to the excited metastable states. The number of electrons raised to such states is proportional to the radiation exposure. This metastable state is maintained until the material is heated at which the electrons come back to the ground state, releasing energy in the form of light. The light is picked up by a photomultiplier tube and converted into an electronic signal. With appropriate calibration, the radiation dose can be straightaway obtained. LiF and CaF are the two of the popular thermoluminescent materials with LiF being the one widely usually in thermoluminescent dosimeter. The advantages of TLD (specially in the case of LiF) are

1. It is independent of the radiation energy
2. It can cover a wide range of radiation doses from a few mrem to many thousands of rem.
3. It is reusable

Its main disadvantage is that it does not provide a permanent record. Personnel monitoring equipment especially film badge or thermoluminescent dosimeter, shall be worn by all radiation workers during radiography work, and when entering radiation areas. Individuals below the age of 18 are not permitted for any job involving radiation or radioactive materials.

It should be mentioned that while area monitoring instruments such as survey meters record the dose rate (in terms of mR/hr or R/hr), personnel monitoring instruments record the total/accumulated dose received by the individual (in terms of mR or R) over a period of time. If the duration of exposure is known, the exposure can be determined from exposure rate of vice versa.

9.3.4 Control of Radiation Hazard

The three cardinal and time tested principles of radiation hazard control distance, time and shielding and these are separately explained as follows:

9.3.4.1 Distance

The greater the distance from a radiation source, lower is the exposure received. The radiation intensities as a function of distance can be calculated using the inverse square law:

$$\frac{I_1}{I_2} = \frac{d_2^2}{d_1^2} \qquad \text{... (9.3.1)}$$

where I_1 and I_2 are the intensities at distances d_1 and d_2 respectively. This is the simplest way to minimise radiation dose. Practically while handling radioisotopes, the distance can be increased by the use of long tongs and remote handling devices. In the case of X- radiography the cables between the control consoles and the X-ray machine should be long enough. The inverse square law is also used to calculate the cordoning distances in case of field radiography. One point to be kept in mind is that the intensities and dose rate calculated as per inverse square law is not accurate, since the radiation intensity at any point is the sum of the direct primary radiation and secondary scatter radiation. Only under ideal conditions of no scatter, the calculated intensities would be exact.

Larger the distance between the source and the operator, lesser is the dose received.

9.3.4.2 Time

The total radiation dose received by the radiation worker in a given field of radiation is directly proportional to the total time spent by the radiation worker in the field.

TLD based on the principle of thermoluminescence

Most widely used crystals are LiF and CaF.

Measures total dose received

Three cardinal principles of hazards control – time distance and shielding

Radiation follows inverse square law. So be as far away as possible

$$\text{Allowable working time} = \frac{\text{Dose Equivalent (DE) limit (mGy / wk)}}{\text{Dose rate (mGy / h)}}$$

The Dose Equivalent Limit (DEL) should be taken only as a general guideline. Any unnecessary exposure to radiation should be avoided. Receipt of the weekly allowable dose in a single exposure should be avoided. All the radiation exposures should be planned in advance and should be executed in the shortest possible time. This reduction in time can be achieved by (a) proper planning of experiments and distribution of work among workers (b) making sure that all the tools which may be required are on hand and then carrying out the job quickly and (c) wherever possible practice on a "mock up".

Lesser the time of operation, lesser will be the Radiation Dose Received

9.3.4.3 Shielding

If maximum distance and minimum time do not ensure an acceptably low radiation level, it is still possible to reduce the radiation level by shielding the radiation source using appropriate high density material such as lead, concrete. Shielding presents protection to worker by maintaining effective barriers between the radiation source and the worker. The process of reduction of radiation intensity (dose) in a barrier is called attenuation. The most common materials used for radiation shielding are lead, concrete, loaded concrete and depleted uranium etc.

Shielding thickness calculations

The required shielding thickness can be calculated from attenuation graphs. These graphs are semilog plots between the reduction factor and the shielding thickness. The reduction factor is defined as the ratio of the radiation level without shield to that with shield.

The required shielding thickness can also be calculated by making use of the half-value thickness or layer and tenth value thickness or layer.

The reduction factor offered by one HVT of shielding material is 2 and by 2 HVT is 2 x 2 or 2^2. The reduction factor offered by 'n' HVT of the shielding material is 2^n.

The reduction factor offered by one TVT is 10 and by two TVT is 10 x 10 or 10^2. The reduction factor offered by 'n' TVT is 10^n. The HVT and TVT values for X-Rays and gamma rays are given in Section - 2.

> Spend as less time as possible in any radiation area.
>
> *This is possible through careful planning, work distribution and mockup.*
>
> *Use appropriate shielding to reduce dose levels.*
>
> *Commonly used shielding materials – lead, concrete, depleted uranium*

Chapter 9.4

Safety in Transport of Radioactive Material

9.4.1 Objectives of Transport Regulation

Industrial radiography using gamma radiography is widely practiced in the country. The radioisotopes for gamma radiography is produced and marketed by the Board of Radiation and Isotope Technology (BRIT), Mumbai. Only qualified, specially trained and experienced persons are authorized to procure and handle radioactive materials. The radioactive materials are transported to the authorized users by various modes of transport. The user returns the decayed sealed sources to BRIT for safe disposal. Transportation of radioactive materials is to be done with utmost care. The packaging should withstand the rigours of transport. In India, Chairman, Atomic Energy Regulatory Board (AERB) is the Competent Authority appointed by Central Government for enforcing regulations relating to radiation safety including transport. Transport of radioactive materials is governed by the surveillance procedures issued under the Radiation Protection Rules – 1971 and the safety code AERB/SC/TR-1, 1986 issued by AERB. This code is based on the international standard of International Atomic Energy Agency (IAEA).

The objective of the recommended regulations for the safe transport of radioactive materials is to protect the public, transport workers, and property from both the direct and indirect effects of radiation during transport. Protection is achieved by limiting the nature and activity of the radioactive material which may be transported in a package of a given design, specifying design criteria for each type of package, and recommending simple rules for handling and stowage during transport.

9.4.2 Packages

Radioactive materials are allowed to be transported only in packages, which have been designed to the prescribed standards. "Package" shall mean the packaging together with its radioactive contents as presented for transport. There are two conditions of transport viz., NORMAL and ACCIDENT. In the NORMAL conditions of transport, packages may be subjected to rough handling, rain and minor mishaps. Under ACCIDENT conditions, it is assumed that the vehicle carrying radioactive material may collide head-on with another vehicle and then engulfed in fire. The type of packaging used to transport radioactive material obviously depends on the nature of the material itself and its activity or quantity. In this context, two terms need to be explained with regard to the physical form of the radioactive material that is to be transported. A radioactive material is stated to be in an 'unsealed form' or 'normal form' if it is in a readily dispersible state (solid, liquid, gas, friable solid etc.) and hence has potential for both internal and external exposure. On the other hand, the material is stated to be in a 'sealed form' or 'special form' (indispersible or sealed capsule), if it is not in a readily dispersible state and hence has potential for external exposure only. The regulatory requirements clearly dictate the type of packaging to be chosen. There are five classes of packages:

- Excepted Packages
- Industrial Packages (Types IP-1, IP-2 and IP-3)
- Type A Packages (Types A_1 and A_2)
- Type B (U) / B (M) Packages (U-unilateral; M-multilateral)
- Type C Packages

Excepted packages are those which contain very small quantities of radioactive material and therefore are excepted from transport regulations. An accident involving such packages will not cause a radiation emergency in the public domain. The AERB guidelines require only proper documentation with respect to these packages. For example, smoke detectors, radio immuno assay (RIA) kits and fluorescent lamp starters are transported in 'excepted' packages.

Industrial packages are used on a wide scale. They have a simple design and are meant for transporting radioactive materials with Low Specific Activity (LSA) and Surface Contaminated Objects (SCO), which are considered intrinsically safe – for example, natural Uranium or Thorium, and tritiated heavy water.

Type A packages are the ones most commonly used (about 75 to 80%). They are designed to be safe enough to withstand the NORMAL conditions of transport but may fail under ACCIDENT conditions. The quantity and type of radioactive material permitted to be transported in these packages are limited to the extent that even if these packages are involved in an accident, any possible release of radioactivity / radiation in the public domain would be well within the regulatory limits. The samples of these packages are actually tested out by simulating the NORMAL conditions of transport. These tests include water spray test, a free drop test, stacking test and a penetration test. Examples of these packages are those used to transport radiopharmaceuticals and nucleonic gauges.

Type B (U) / B (M) packagesare designed to be safe enough to withstand under NORMAL as well as ACCIDENT conditions arising during its transport. The samples of these packages are actually tested out by simulating accident conditions to demonstrate their integrity. For example, they are subject to various tests viz., 1) a drop test (dropping the package from a height of 9 meters on to a hard unyielding surface and followed by dropping it on a mild steel bar from a height of 1 meter), 2) a fire test (subjecting the package to a temperature of 800 degrees Celsius for 30 minutes) and 3) a water immersion test (subjecting the package to an external pressure equivalent to 15 or 200 meters depth of water). The difference between type B (U) and B (M) packages essentially pertains to their certification for international use. Type B (U) certification by the AERB would be accepted internationally and type B (M) certification would need to be revalidated by the appropriate regulatory authority of the concerned country through which the package would transit.

Type C packages are designed to withstand very severe accident conditions characteristic of transport by air. The criteria for testing these packages include very high impact test (equivalent to velocities of 90 meters per second), enhanced fire tests (say at 800 degrees Celsius for 60 minutes) and puncture/tearing tests.

9.4.3 Labelling and Marking

The radioactive source should be transported only in a shielded container, which has been previously approved by the Competent Authority. The source must be fastened within the shielded container by means of appropriate locking mechanism incorporated in the design of the shielded container.

The lid or shutter of the shielded container must be securely closed and locked so that the source would not be released during the transport. The shielded container must be loaded in an outer container such as a sturdy wooden box provided with spacers within for preventing free movement of the shielded container inside, during transport. It must be ensured that the outer container deployed for housing the loaded shielded container is in sound condition. The outer container must be provided with strong lifting handles in order to facilitate handling of the loaded container. The outer container must be closed and locked. The outer container housing the shielded container duly incorporating the source is referred to as the package.

The radiation level on the external surface of the package must be monitored using an appropriate radiation survey meter properly calibrated and in good working condition. The maximum radiation level at the external surface of the package should be recorded. If it is observed that this radiation level is zero or very low, then the observation must be counter-checked by means of another radiation survey meter which is known to be in good working condition and by appropriately making sure that the source is contained within. In the case of teleflex type cameras care must be taken to ensure that the source is in the safe position within the shielded container and that it does not remain in the guide tube during transport. The camera must be securely locked to prevent movement of the source during transport.

Transport index for a package shall mean the number expressing the maximum radiation level in mrem/h at one meter from the external surface of the package. The transport index of the package must be determined by means of an appropriate radiation survey meter in good working condition and this transport index must be mentioned clearly along with other details on the package. It is the label affixed on the package, which communicates with the outside world about the nature of the consignment. The presence of a label makes it clear to the onlooker that the package contains a material, which may be special. The legend RADIOACTIVE written on the label clearly announces the exact nature of the contents. For those who may not be able to read the inscription, the radioactive symbol would indicate the nature of consignment. It is helpful to know the symbols, which represent the various materials such as explosives, inflammables, corrosives and poisons. The labels, affixed on the exterior of the package, furnish the following particulars regarding the package:

- Contents are radioactive
- The name of the radioactive source and its activity
- Type of the package
- Category of the package
- Transport index
- Gross weight of the package
- Addresses of sender and receiver

A package may belong to one of the following categories:

- Category I – WHITE label
- Category II – YELLOW label
- Category III – YELLOW label

The total sum of the transport indices of the packages in any one large freight container shall not normally exceed 50. Transport index and the maximum radiation level at any point on external surface of the packages for the above mentioned categories of packages are given in Table 7.1 below. The appropriate category label must be affixed atleast on the two opposite sides of the exterior of each package (The labels for the three categories and the placard for vehicles is given in Figs. 9.4.2 – 9.4.5). As mentioned earlier, the addresses of the consignor and the consignee must be clearly and unambiguously inscribed on the exterior of the package. The package must not be transported as personal luggage. The package must always be booked as a cargo by road, rail, air or water. The package must be declared as a radioactive consignment UN class 7 in the transport documents.

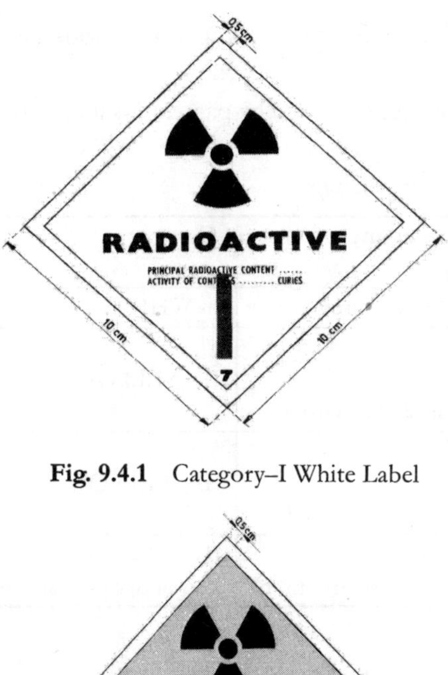

Fig. 9.4.1 Category–I White Label

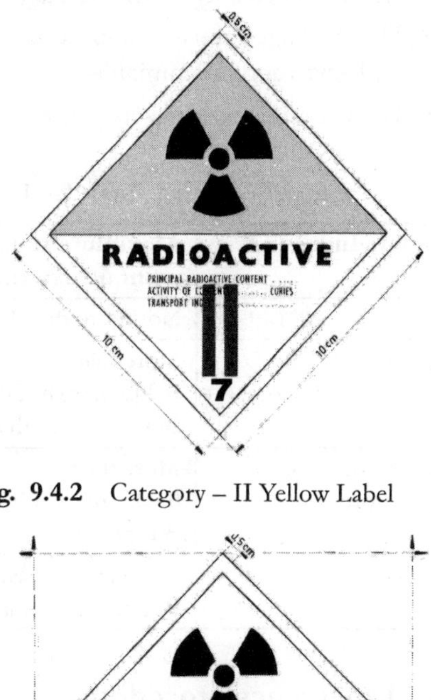

Fig. 9.4.2 Category – II Yellow Label

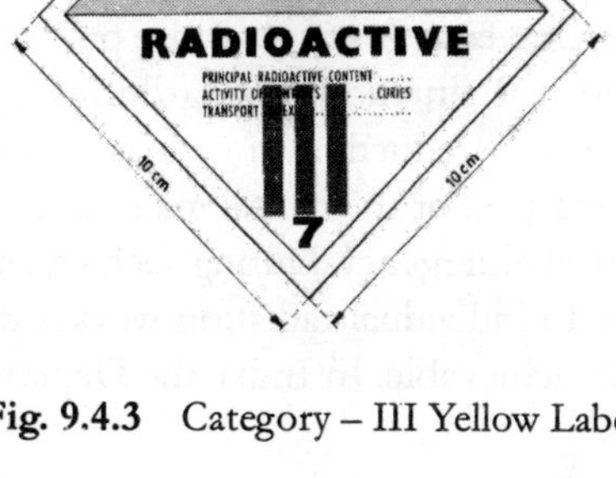

Fig. 9.4.3 Category – III Yellow Label

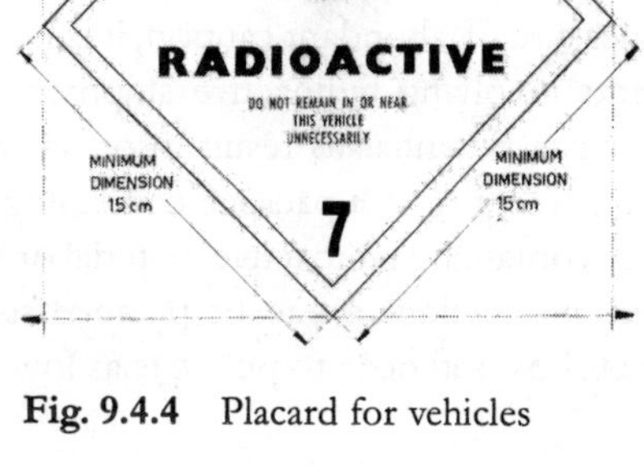

Fig. 9.4.4 Placard for vehicles

It may be seen from the different category labels and placard for vehicle that number 7 is inscribed as per the classification of dangerous goods by UN. The nine classes as determined by the UN Committee of experts on the transport of dangerous goods, are given in the Annexure 7.1. While booking the cargo, a certificate must be issued by the consignor to the effect that the package containing a radioactive material has been properly packed, marked and labeled and the consignment is in conformity with the requirements for the safe transport of radioactive materials. The consignee must be duly informed and it must be ensured that the consignee would be willing to clear the consignment (prior to dispatch of the package to any destination). The carrier shall be instructed as follows:

- The package must be transported by the direct route
- Intermediate off-loading must be avoided
- The package must be handled carefully; fork lifts/cranes/pallet trucks should be used, if package weighs more than 30 kg
- Persons must not be allowed to sit on the package or to spend more than the necessary time in the vicinity of the package
- The packages must not be transported along with other dangerous goods such as explosives and inflammables
- The package must be kept segregated from spaces occupied by passengers and public

Table 9.4.1 Categories of packages

Transport Index (TI)	Maximum radiation levels at any point on external surface	Category
Zero (0)	Not more than 0.005 mSv/h (0.5 mrem/h)	I - WHITE
more than zero (0), but not more than one (1)	More than 0.005 mSv/h (50 mrem/h) But not more than 2 mSv/h (200 mrem/h)	II - YELLOW
more than one (1), but not more than 10	More than 0.005 mSv/h (0.5 mrem/h) But not more than 0.5 mSv/h (50 mrem/h)	III - YELLOW
more than 10	More than 2 mSv/h (200 mrem/h) But not more than 10 mSv/h (1000 mrem/h)	III - YELLOW And also under exclusive use

9.4.4 Emergency procedure

As a measure of abundant caution, it is necessary to develop emergency response procedures for accidents involving radioactive shipments. Any off-normal situation arising during transport which has a potential to result in or has actually resulted in reduction in or loss of either the shielding integrity of a package containing radioactive material or the containment integrity of a package containing radioactive material or both is called an emergency. During such emergencies, the response actions are to be planned such that dose to individual radiation worker does not exceed 0.1 Sv and dose to public is as low as reasonably achievable. In India, the Department of

Atomic Energy (DAE) has appointed the Crisis Management Group (CMG) for co-ordinating the response efforts in the event of any off-site radiological emergency including those involving radioactive shipments.

Questions

Choose the Most Appropriate Answer

1. Exposure to X-rays or gamma rays
   a. may have a cumulative effect which must be considered
   b. will be beneficial since they build up an immunity to radiation poisoning
   c. will have no effect on human beings
   d. will have only a short-term effect on human tissues
   e. the effect cannot be predicted
2. Gamma sources with high activity are best handled using
   a. special protective apron and clothing
   b. remote handling equipment
   c. spending very less time near the radiography source
   d. minimising the radiographic exposure time so that personnel exposure is also reduced
   e. no specific precautions are necessary
3. Of the following which would be especially sensitive to injury by excessive exposure to X- or gamma rays?
   a. blood
   b. lens of the eye
   c. internal organs
   d. brain
   e. all of the above
4. A piece of lead having a length equivalent to 4 HVT and thickness equivalent to 1 HVT is placed in the path of a beam of radiation from a Co-60 source. It will reduce the dose rate at a given location by
   a. 25 %
   b. 50 %
   c. 75 %
   d. 93.75 %
   e. 96 %
   f. inadequate information
7. The exposure of personnel to X-and gamma radiation can be measured or monitored by means of:
   a. film badges
   b. dosimeters
   c. radiation exposure survey meters
   d. all of the above
8. An advantage of the pocket dosimeter type of ionization chamber used to monitor radiation received by personnel is:
   a. it provides a permanent record of accumulated dosage
   b. it provides an immediate indication of dosage
   c. it is the most sensitive detector available
   d. all of the are advantages

9. In making an isotope exposure in an unshielded area. You find the dose rate 1.8m (6 ft) from the source is 1.200 mR/h. What would be the dose rate at 7.3 m (24 ft)?
   a. 75 mR/h
   b. 100 mR/h
   c. 200mR/h
   d. 300 mR/h
10. The intensity of X or gamma radiation is measured in
    a. roentgens
    b. ergs
    c. roentgens per unit of time
    d. H & D units
11. An unshielded isotope source gives a dosage rage of 900 mR/h at 3m (10 ft.). What would the unshielded dosage rate be at 9m (30ft)?
    a. 300 mR/h
    b. 600 mR/h
    c. 100 mR/h
    d. 2.700 mR/h
12. A radioactive source with an activity of 1 Ci has
    a. 1000 disintegrations per second taking place
    b. 1000 000 disintegrations per second taking place
    c. 1000 000 000 disintegrations per second taking place
    e 3.7×10^{10} disintegrations per second taking place
13. The most common material used to provide protection against X-rays is
    a. high density brick
    b. an alloy of 70 percent Steel and 30 percent Copper
    c. tungsten
    d. lead
14. Unit of radioactivity is:
    (a) rad
    (b) rem
    (c) Sievert
    (d) Becquerrel
15. Annual effective dose limit for occupational workers is:
    (a) 1 mSv
    (b) 10 mSv
    (c) 20 mSv
    (d) 30 mSv
    (e) 50 mSv

SECTION 10
ADVANCES IN RADIOGRAPHY

After reading this section, you should be able to answer

★ *What is microfocus, real time and high energy radiography*

★ *Principles, advantages and typical applications of these techniques*

Chapter 10.1

Microfocal Radiography

10.1.1 What is Microfocal Radiography?

The technique of radiography is based on the principles of differential absorption and shadow projection. Thus, one of the main parameters which determine the image quality and the detectability of features is the area from where the X-rays originate commonly known as the "focal spot". The effect of focal spot on radiographic sensitivity has been discussed in earlier sections. It can be observed from Fig. 10.1.1 that while a point source of radiation produces a sharp image, a large source produces a diffused image with a large geometric unsharpness. This geometric unsharpness U_g primarily depends on the focal spot size as given by the relation

$$U_g = \frac{f \times OFD}{SOD} \quad \text{... (10.1.1)}$$

where OFD is the object to film distance and SOD the source to object distance. In conventional radiography units, the size of the focal spot ranges from 1 to 5 mm. Hence to keep U_g as low as possible, the film is placed in intimate contact with the object (minimizing OFD) and the source to object distance is increased. However, the SOD cannot be increased beyond a limit, since this would make the exposure times impractical. An alternative method is to reduce the focal spot. X-ray equipment in which the size of the focal spot is between 0.1–1 mm, is commonly referred to as minifocus unit while X-ray equipment in which the focal spot size is less than 0. 1 mm or 100 micrometers is referred to as microfocus unit. This small focal spot is achieved by focussing the electron beam on to the target. Present day microfocus units have focal spots in the range of 5 – 15 micrometers. The fundamental physical processes such as the electron scattering in the target makes it difficult to achieve focal spots better than 5 micrometers.

10.1.2 Advantages and Limitations

Once the focal spot size is reduced, a number of advantages can be identified. These include:

(a) ***Projection magnification*** : The object need not be in contact with the film during exposure as in conventional radiography. Thus one can obtain enlarged primary radiographs with magnifications greater than 2X (Fig. 10.1.2). Magnification reduces the number of features that is masked by the background image noise thus enhancing the detection sensitivity of microdefects.

> Microfocus unit -
> *X-ray equipment in which focal spot is less than 100 micrometers*
>
> Smaller focal spot makes possible
> *Projective magnification without loss of definition*
> *Improves radiographic image defintion*
>
> Microfocal radiographs are excellent in contrast and sensitivity

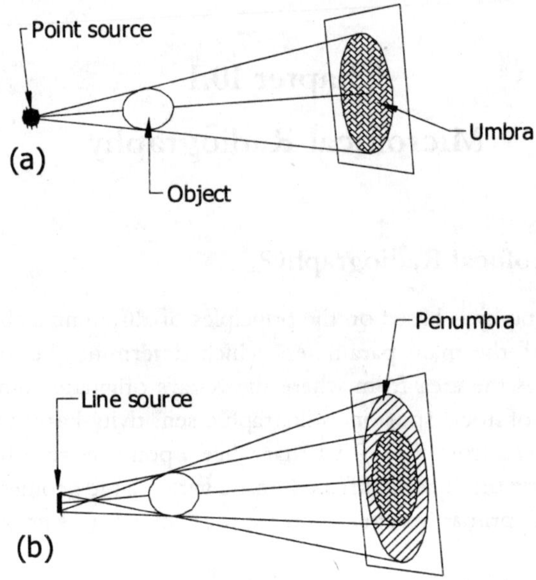

Fig. 10.1.1 Image due to Point source of radiation and line source of radiation

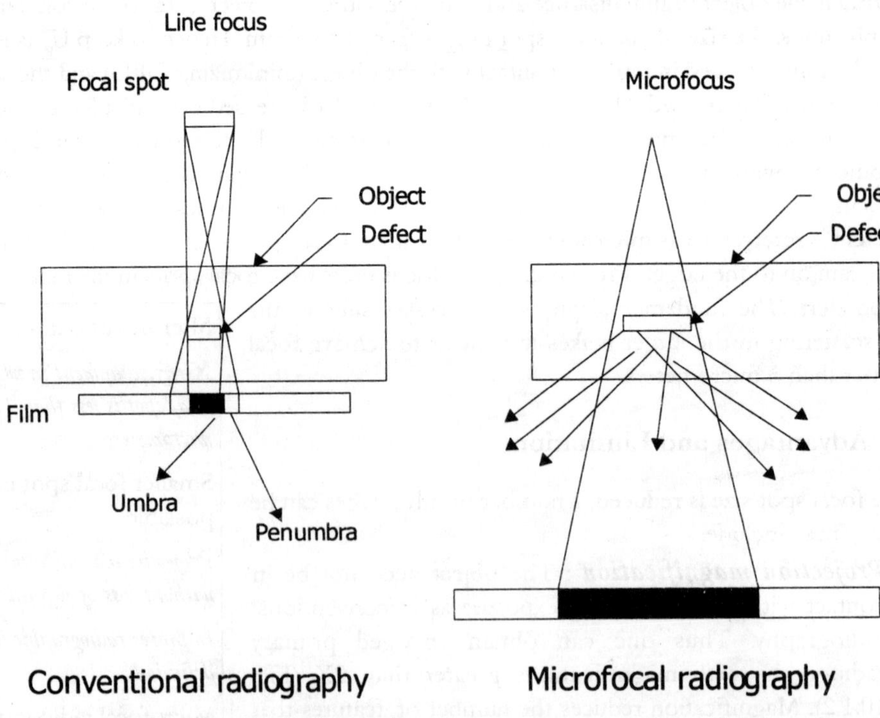

Fig. 10.1.2 Microfocal spot makes it possible to have projective magnification and high contrast radiographs due to decreased scatter radiation.

(b) ***Improved radiographic contrast*** : It is well known that in conventional radiography, scattered radiation especially generated from within the object reduces radiographic contrast to the maximum. Once the object is placed away from the film, the amount of scattered radiation reaching it is drastically reduced. Thus, microfocal radiographs have much better contrast compared to conventional radiographs.

(c) ***Possibility of object manipulation*** : Since the object and the film can be separated without sacrificing image definition, real time radiography of dynamic/temporally changing events is possible. Further the object can be rotated / translated within the radiation beam making stereo and micro tomography possible. These techniques allow better detection of planar defects and greater resolution of detail within the section thickness.

Limitations

Projection magnification has its inherent disadvantages.

(a) Since the object is placed closed to the source, a smaller volume of the object is inspected at any one time as can be seen from Fig. 10.1.1. This means more number of exposures and more number of films.

(b) Since the electrons are focussed on to the target, the heat is concentrated in a very small and localised spot. Hence the target cannot be loaded to a great extent which limits the tube current.

However, both of these are not very serious limitations.

10.1.3 Applications

The main areas of application of microfocal radiography can be classified as:

(a) those where conventional radiography cannot be applied due to problems of access such as evaluation of tube to tubesheet welds of steam generator, mode couplers etc.

(b) those where conventional radiography can be applied but cannot resolve the fine defects necessary to be detected example

(i) detection of microdefects such as voids, microcracks and inhomogenous distribution of material in ceramics and

(ii) detection of defects in microelectronic components

(c) for real time and computed tomography applications where microfocal radiography is an essential requirement due to higher unsharpness of the real time systems example online evaluation of automotive components.

Chapter 10.2
Fluoroscopy and Real Time Radiography

Real time radiography (RTR) or fluoroscopy differs from conventional radiography in that the X-ray image of the object is observed directly on a screen rather than recorded on a film.

10.2.1 Fluoroscopy

In the earlier days, fluorescents screens such as Zinc Sulphide or Calcium Tungstate was used. The X-ray beam after passing through the object falls on the fluorescent screen. The X-rays are converted to light photons and the radiographic image could be directly viewed. Fluoroscopic systems had the advantage of high speed and low cost. The main disadvantages of these systems were

(a) only low energies could be used (due to problems of radiation exposure), the range of thickness that could be examined was also low.

(b) the high screen unsharpness of the fluorescent screens resulted in poor radiographic contrast and sensitivity

(c) the low brightness of the images led to operator fatigue quite quickly and

(d) there was a lack of permanent record.

10.2.2 Real Time Radioscopy

The advent of image intensifier tubes in the 1950s led to the development of real time radioscopy systems which overcame all the problems of fluoroscopic systems. Present day real time systems use image intensifiers, video camera and monitor. The image intensifier converts X-ray photons to electrons, accelerates the electrons and then reconverts them to light which is picked up by the CCD camera and displayed on a monitor. Figure 10.2.1 shows a block diagram of a typical RTR setup.

10.2.2.1 Principle of Operation

The RTR system essentially consists of an image intensifier tube, a zoom lens, CCD camera, image processor and a monitor. When a beam of X-rays is incident on the input screen of the image intensifier, the X-ray photons are converted into light photons by the detector – Caesium iodide. A photocathode converts the light photons to electrons which are accelerated and impinge on the output phosphor. The output phosphor converts the electrons to light photons which are picked up by the CCD camera and after image processing displayed on the TV monitor.

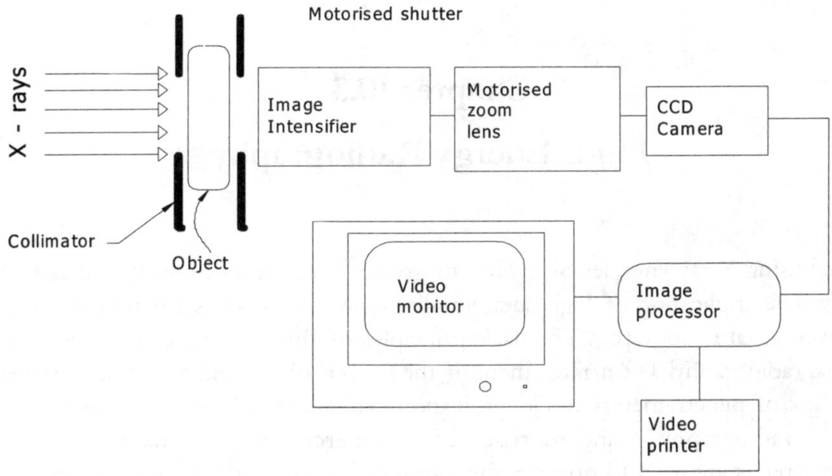

Fig. 10.2.1 Block Diagram of real time radiography setup

10.2.3 Advantages, Limitations and Application Areas

Improvements in electronic gain, fluorescent and photo cathode layer efficiency and electron optics have made modern X-ray image intensifiers very useful in industrial applications. The main advantages of RTR include an overall savings both with respect to man-hours spent, cost and archiving the radiographic image. Accept-or-reject decisions can be made immediately without the delay in film development. This has been found to be cost effective in the long run.

Limitations of RTR are its higher inherent screen unsharpness, which limits the overall radiographic sensitivity as compared to that of film radiography. However, through the use of image processing, the sensitivity levels can be improved to those obtainable in film radiography.

RTR is presently applied to objects on assembly lines for rapid inspection. It finds extensive applications in real time evaluation of automotive components such as wheel castings, valves, and online inspection of seam welds, canned foods and in a variety of other industries. Fig. 10.2.2 shows a typical real time image of a weld with lack of penetration and porosities.s

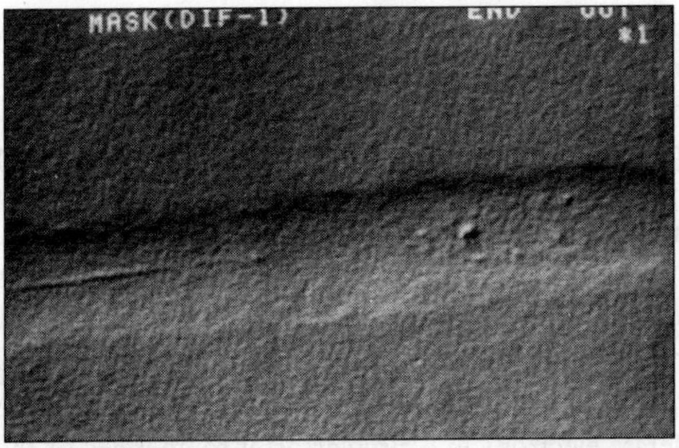

Real time radiography

Based on image intensifier systems

Provides online evaluation and image archiving

Cost effective

Used for automated inspections

Fig. 10.2.2 Real time image of weld after image processing and edge enhancement indicating lack of penetration and porosities.

Chapter 10.3

High Energy Radiography

Radiography using X-ray energies of 1 MeV or greater is commonly considered as high energy radiography. The application of high energy radiography for NDT is a natural outcome of the use of conventional radiography. The basic principles of this technique are similar to those of conventional radiography. The arrangement of the source, object and film, the scatter reduction techniques, use of penetrameters and identification markers are all similar to those adopted with conventional radiography. Standard types of commercial X-ray films with lead or other intensifying screens are used to produce the radiographic image of the object being examined. The difference between high and low energy radiography arise from several distinctive characteristics of a high energy X- radiography source many of which prove to be advantageous.

10.3.1 High Energy Radiography Sources

The main types of high energy sources include: -

(a) ***Resonant transformer:*** The 1 MeV resonant transformer was the first commercial high energy X-ray source introduced by the General Electric of USA in 1939. A few years later, a 2 MeV version of the Resotron was fabricated. This kind of unit operates on the basis of a coupling between tuned primary and secondary windings and a sectionalized X-ray tube. The electrons are accelerated by application of the full d-c potential across the tube. These units are seldom used today.

(b) ***Van De Graff Generator:*** Though conceived earlier, the Van De Graff style electrostatic generator also became commercially available around 1939 in the 1 and 2 MeV range. In this unit, electric charges obtained from a d-c potential of about 30kV are transported by means of a moving belt to the high potential electrode. The full potential of 1 to 2 MeV is applied across a sectionalized X-ray tube.

(c) ***Linear Accelerator:*** Also popularly called as "Linacs", these are machines that are based on the wave guide technique. A high-frequency wave passes down a hollow conducting tube. Electrons injected into the tube at about 50kV are accelerated to energies of the order of 3 to 40 MeV, by transfer of energy from the high-frequency wave. Apart from X-radiographic applications, Linacs can also be used as a source of neutrons by the addition of a Uranium Be target. Of all the high energy equipments, Linac is the most widely used high energy source for radiography. With linacs, radiography of steel upto 500 mm(20 inches) is possible. They will produce better radiographs than almost any other equipment for steel thicker than 100mm.

(d) ***Betatron:*** This is basically a combination of an electromagnet and a transformer. It is designed to guide and accelerate electrons in a circular orbit to energies of the order of 15 to 10 MeV using an electrical field induced by a time-varying magnetic field. The beam currents are very small of the order of few microamperes while the focal spot is less than 1 mm. The small focal spot and highly collimated beam makes it possible to obtain radiographs with very high contrast and sensitivity (better than 1%). Steel thickness that can be radiographed using these machines range from 50 mm - ~400 mm.

10.3.2 Advantages of High Energy Radiography

The major advantages of high-energy radiography include

(a) It is possible to examine thicker sections economically due to the greater penetration by the higher energy photons.

(b) Large distance over thickness ratios (D/T) can be used with correspondingly low geometrical distortion.

(c) Short exposure times and high production rates are possible.

(d) The wide thickness latitude, good contrast, reduced amount of high angle scatter reaching the film, small focal spot sizes (1 - 2 mm) results in radiographs with excellent sensitivity, contrast and detail resolution.

Thus high energy radiography is an attractive choice for the inspection of thick metal castings, weldments and examination of high density materials like Uranium alloys etc. However, the overall cost of the equipment is quite high compared to conventional radiography.

Summary

Radiography using radiation energies > 1 MeV called high energy radiography

Sources for high energy radiography include

* *Resonant transformers*
* *Van De Graff generators*
* *Betatrons*
* *Linear accelerators (most widely used)*

Advantages of high energy

* *Wide thickness latitude can be examined*
* *Short exposure times*
* *Excellent contrast and sensitivity*

General References

1. Halmshaw. R, Physics of Industrial Radiology, Heywood books, London, 1966.
2. Halmshaw, R., Mathematics and Formulae in NDT, British Institute of Nondestructive Testing, 1983.
3. Halmshaw. R, Industrial Radiology Techniques, Wykeham Publication Ltd, London, 1971.
4. Halmshaw. R, Industrial Radiology: Theory and Practice, Applied Science Publishers, London, 1982.
5. Halmshaw. R , Non-Destructive Testing, Edward Arnold Publishers, 1987.
6. Halmshaw. R , Introduction to the Non-Destructive Testing of Welded Joints, Abington , 1988.
7. Herz R.H., The Photographic Action of Ionizing Radiation, Wiley Inter science, New York, 1969.
8. Agfa-Gevaet N.V., Industrial Radiography, Morstel, Belgium, 1992.
9. Nondestructive Testing Handbook Ed 3 Vol.3: Radiography and Radiation Testing, Bryant, L E & Mclntire, American Society for Nondestructive Testing, 1988.
10. International Institute of Welding, Handbook on Radiographic Apparatus and Techniques, International Institute of Welding, 1961.
11. Ely. R, Microfocal Radiography, Academic Publishers. UK, 1982
12. Neutron Radiography Handbook Ed. P. Vander Hert and H. Rottger, D. Riedel Publishing Co., Holland, 1981.
13. Venkataraman.B, Baldev Raj, Viswanathan.K, Neutron Radiography – Principles and Applications, Neutron Radiography Working Group of India, 1999.
14. Practical Applications of Neutron Radiography and Gaging, ASTM STP 586. 1975.
15. Baldev Raj, Jayakumar.T and Thavasimuthu.M, Practical Non- Destructive Testing, Narosa Publishers, New Delhi, 1996.
16. Richard Quinn. A, Claire Signal. C, Radiography In Modern Industry Ed 4, Eastman Kodak Company, France, 1980.
17. Classroom Training Handbook - Radiographic Testing, General Dynamics, Convair Division, San Diego, California, 1967.
18. Programmed Instruction Handbook, Non-Destructive Testing - Radiographic Testing, General Dynamics, Convair Division, San Diego, California, 1967.
19. Justin. G. Schneeman, Industrial X-ray interpretation, Intex Publishing Company, Evanston, 1968.
20. Don. E Bray, Roderick. Stanlely, Non Destructive Evaluation, Mc Graw-Hill Book company, New York, 1989.

21. Subramaniam. C.V., Baldev Raj and Jayakumar. T, Non Destructive Testing of Welds, Narosa Publishers, New Delhi. 2000.
22. ASM International Handbook, Vol.17, Non Destructive Evaluation and Quality control, ASM International, U.S.A, 1989.
23. Radiography of Welds, International Institute of Welding, Sweden, 1968.
24. Mix. P.E., Introduction to Nondestructive Testing: Training Guide, John Wiley Publishers, New York, 1987.
25. Industrial Radiography, Training Guidelines in Non-destructive Testing Techniques, International Atomic Energy Agency, Vienna, 1992.

Web References

1. http://www.ndt-ed.org/EducationResources/CommunityCollege/Radiography/Physics.
2. http://www.kodak.com/global/en/service/pubs/kpro/radiography

Index